Youth and the Rural Economy in Africa: Hard Work and Hazard

Youth and the Rural Economy in Africa: Hard Work and Hazard

Edited by

James Sumberg

Institute of Development Studies (IDS), Brighton, UK

CABI

CABI is a trading name of CAB International

CABI
Nosworthy Way
Wallingford
Oxfordshire OX10 8DE
UK

Tel: +44 (0)1491 832111
Fax: +44 (0)1491 833508
E-mail: info@cabi.org
Website: www.cabi.org

CABI
WeWork
One Lincoln Street
24th Floor
Boston, MA 02111
USA

Tel: +1 (617)682-9015
E-mail: cabi-nao@cabi.org

A catalogue record for this book is available from the British Library, London, UK.

Library of Congress Cataloging-in-Publication Data

Names: Sumberg, J. E. editor.
Title: Youth and the rural economy in Africa : hard work and hazard / James Sumberg.
Description: Boston, MA : CAB International, [2021] | Includes bibliographical references and index. | Summary: "There are many accepted views about the role and position of youth in the African agricultural economy. This book challenges those assumptions by analysing new studies, particularly of gender and social difference. It is a major contribution to current debates and development policy about youth, agriculture and employment in rural Africa"-- Provided by publisher.
Identifiers: LCCN 2021005799 (print) | LCCN 2021005800 (ebook) | ISBN 9781789245011 (hardback) | ISBN 9781789249828 (paperback) | ISBN 9781789245028 (pdf) | ISBN 9781789245035 (epub)
Subjects: LCSH: Agriculture--Economic aspects--Africa. | Youth--Africa. |Rural development--Africa. | Africa--Rural conditions.
Classification: LCC HD2116 .Y688 2021 (print) | LCC HD2116 (ebook) | DDC 331.3/83096--dc23
LC record available at https://lccn.loc.gov/2021005799
LC ebook record available at https://lccn.loc.gov/2021005800

References to Internet websites (URLs) were accurate at the time of writing.

ISBN-13: 978 1 78924 501 1 (hardback)
978 1 78924 502 8 (ePDF)
978 1 78924 503 5 (ePub)

DOI: 10.1079/9781789245011.0000

Commissioning Editor: David Hemming
Editorial Assistant: Emma McCann
Production Editor: Marta Patiño

Typeset by SPi, Pondicherry, India
Printed and bound in the UK by Severn, Gloucester

Contents

Contributors

Chamberlin, Jordan, International Maize and Wheat Improvement Center (CIMMYT), Nairobi, Kenya. E-mail: j.chamberlin@cgiar.org
Crossouard, Barbara, University of Sussex, Brighton, UK. E-mail: b.crossouard@sussex.ac.uk
Dunne, Máiréad, University of Sussex, Brighton, UK. E-mail: mairead.dunne@sussex.ac.uk
Flynn, Justin, Institute of Development Studies (IDS), Brighton, UK. E-mail: j.flynn2@ids.ac.uk
Oosterom, Marjoke, Institute of Development Studies (IDS), Brighton, UK. E-mail: m.oosterom@ids.ac.uk
Sumberg, James, Institute of Development Studies (IDS), Brighton, UK. E-mail: j.sumberg@emeritus.ids.ac.uk
Szyp, Carolina, Institute of Development Studies (IDS), Brighton, UK. E-mail: c.szyp1@ids.ac.uk
Thorsen, Dorte, Institute of Development Studies (IDS), Brighton, UK. E-mail: d.thorsen@ids.ac.uk
Yeboah, Felix Kwame, Michigan State University (MSU), East Lansing, MI, USA. E-mail: yeboahfe@msu.edu
Yeboah, Thomas, Bureau of Integrated Rural Development (BIRD), Kwame Nkrumah University of Science and Technology (KNUST), Kumasi, Ghana. E-mail: thomas.yeboah@knust.edu.gh

Acknowledgements

In addition to the chapter authors, many other individuals contributed to *Youth and the Rural Economy in Africa: Hard Work and Hazard.*

First and foremost, we are deeply grateful to the hundreds of young people, and others, who took the time to participate in the research.

Field work was expertly led by Dr Victoria Flavia Namuggala (Makerere University, Uganda), Prof. Okello Uma Ipolto (Gulu University, Uganda), Dr Mamo Hebo Wabe (Addis Ababa University, Ethiopia), Dr Fekadu Adugna Tufa (Addis Ababa University, Ethiopia), Dr Bela Teeken (International Institute for Tropical Agriculture, Nigeria), Dr Tessy Madu (National Root Crops Research Institute, Nigeria) and Dr Affoué Philomène Koffi (Université Félix Houphouët-Boigny, Côte d'Ivoire). In each country these individuals worked with one or more teams of field researchers, and their dedication throughout the research process is very gratefully acknowledged.

Dr Hailemariam Ayalew, Dr Kibrom Abay and Ms Woinishet Asnake made valuable contributions to the analysis of LSMS–ISA data.

Prof. Agnes Andersson Djurfeldt, Dr Keetie Roelen and Prof. Rachel Sabates-Wheeler provided valuable reviews of draft chapters.

The research was undertaken with grants from the International Fund for Agricultural Development (IFAD), the Agricultural Policy Research in Africa programme (APRA) (funded by the UK Department for International Development, DFID), and the MasterCard Foundation; their support is gratefully acknowledged. The opinions expressed here belong to the authors, and do not necessarily reflect the views or policies of these funders.

Finally, without the support and critical eye of Dr Christine Okali, this book would never have seen the light of day.

Abbreviations

AGRA	Alliance for a Green Revolution in Africa
APRA	Agricultural Policy Research in Africa programme
BEPC	*Brevet d'Etudes du Premier Cycle*
BIRD	Bureau of Integrated Rural Development, Kumasi, Ghana
CIMMYT	International Maize and Wheat Improvement Center, Nairobi, Kenya
CSO	Zambian Central Statistical Office
DFID	UK Department for International Development
FGD	focus group discussion
FTLRP	Fast Track Land Reform Programme
GDP	gross domestic product
HCT	human capital theory
IAPRI	Indaba Agricultural Policy Research Institute
ICTs	information and communications technologies
IDP	internally displaced persons
IDS	Institute of Development Studies, Brighton, UK
IFAD	International Fund for Agricultural Development
ILO	International Labour Organization
IMF	International Monetary Fund
IT	information technology
KNUST	Kwame Nkrumah University of Science and Technology, Kumasi, Ghana
LRA	Lord's Resistance Army
LSMS–ISA	Living Standards Measurement Study – Integrated Surveys on Agriculture
MDGs	Millennium Development Goals
MSU	Michigan State University
NCE	Nigeria Certificate in Education
NGO	non-governmental organization
PLE	Primary Leaving Exam
RALS	Rural Agricultural Livelihood Survey
SAPs	Structural Adjustment Programmes
SDGs	Sustainable Development Goals
SI	sustainable intensification
SSA	sub-Saharan Africa
SUR	seemingly unrelated regressions
SWTS	School-to-Work-Transition Surveys

1 African Youth and the Rural Economy: Points of Departure

James Sumberg[1], Justin Flynn[1], Marjoke Oosterom[1], Thomas Yeboah[2], Barbara Crossouard[3] and Dorte Thorsen[1]
[1]Institute of Development Studies, Brighton, UK; [2]Bureau of Integrated Rural Development (BIRD), Kwame Nkrumah University of Science and Technology (KNUST), Kumasi, Ghana; [3]University of Sussex, Brighton, UK

Introduction

How do young people across Africa engage with the rural economy? What are the implications of this engagement for their efforts to build their livelihoods, and for their futures, for society and for rural areas? These are the questions that motivate this book and the research that underpins it. Such questions will be of interest to researchers, policy makers, development professionals and others concerned with the well-being and aspirations of young people, with their search for employment and decent work, and with the relationship between schooling and work. Individuals working on rural poverty and food security, agriculture and rural development – and rural transformation more broadly – should certainly be interested in rural young people's lives and livelihoods, and the futures they imagine for themselves. Finally, a more nuanced understanding of young people's engagement with the rural economy can help to ground debates about demographic change, including migration and urbanization, and provide a much needed reality check of common assumptions and narratives concerning youth, conflict and radicalization.

The fact that a number of these same concerns – including education, decent work and migration – are integral to the Sustainable Development Goals (SDGs), and that several of the SDGs speak directly to the situation of youth, demonstrates the central place that young people have come to occupy in development debates and policy. Indeed, there is a growing body of youth-focused scholarship, policy analysis, implementation guidance and programme evaluations – as well as a plethora of youth-targeted development initiatives. Taken together, these suggest that youth in rural Africa are being taken seriously, and it appears that this focus will continue well into the future. Whether they are being taken seriously for the right reasons, and whether they are well served by the policy and development investments made in their name, are important points of debate.

The book's ambition is to advance the understanding of young people as social and economic actors in rural Africa. It does this through new empirical analyses, both quantitative and qualitative, involving a significant number of rural young people across multiple countries. These new analyses are brought to bear on the narratives and debates that frame and channel much of the current interest in youth-specific policy and investment.

At this point, readers might be asking themselves, 'With the recent publication of *Creating Opportunities for Rural Youth* (IFAD, 2019) and *Youth and Jobs in Rural Africa: Beyond Stylized Facts* (Mueller and Thurlow, 2019), do we really

DOI: 10.1079/9781789245011.0001

need another book on African rural youth?' Our response is an emphatic 'Yes', based primarily on the fact that neither of these two works bring the histories, lives, voices or imagined futures of rural youth into the equation. *Youth and the Rural Economy in Africa: Hard Work and Hazard* begins to address this critical lacuna.

To allow the voices of young people to emerge, this book both starts with different questions and draws from an expanded set of intellectual and conceptual traditions, and data sources. For example, the first question Mueller *et al.* (2019) pose in *Youth and Jobs in Rural Africa: Beyond Stylized Facts* is, 'Are rural youth active participants in the national growth process?' They go on to ask how their involvement in agricultural technology adoption, rural income diversification and urban migration 'affect rural transformation' (Mueller and Thurlow, 2019, p.3). It is clear from this that while the book investigates 'the role of rural youth in sub-Saharan Africa's (SSA) development' (Mueller and Thurlow, 2019, p.3), the primary interest is in national growth processes and rural transformation, not youth. This explains the prominence given to Timmer's four-stage model of agricultural transformation (Timmer, 1988) and the striking absence of any theoretical or conceptual treatment of youth as social and economic actors. Mueller and Thurlow (2019) and IFAD (2019) rely almost exclusively on survey data collected through exercises that generally were not designed with a particular youth focus in mind.

In contrast, in *Youth and the Rural Economy in Africa: Hard Work and Hazard* we start with the simple question, 'What are rural young people doing?' In placing their actions, and their views about those actions, at centre stage, we make no assumptions about what they should be doing, how or where they should be doing it, or what their motivations should be. This is not to say that we approached the research without preconceptions or hypotheses – indeed, as will become clear, we draw on a wide array of conceptual insights and disciplinary approaches. While not abandoning microeconomic analytical frameworks and survey data analysis, we have made a conscious effort to bring these together with relevant literature from the broader social sciences including anthropology, sociology, social geography, youth studies, gender studies, education and policy studies, and with a wider range of data and modes of analysis. In so doing we have sought to grapple with the heterogeneity – of rural areas, family contexts and young people – which is still largely overlooked by the majority of policy-oriented analyses.

This chapter proceeds as follows. The next section situates the current interest in Africa's rural youth, and the place of this book, within the broader discussion of policy narratives. It then identifies seven narratives about rural youth in SSA that channel much contemporary policy and development intervention. Following this the argument that runs through the book is outlined. The key conceptual resources that the various chapters draw upon are briefly introduced in the following section. The last section provides a brief summary of each of the subsequent chapters.

Policy Narratives

What they are and why they matter

As with all policy problems, policy and interventions relating to rural youth in SSA are built around narratives or stories (Roe, 1991, 1995; Jones and McBeth, 2010). Narratives are central to policy processes, serving as an important vehicle for organizing and communicating policy information (Shanahan *et al.*, 2011). They set out the problem, explain why it has arisen and propose how it should be addressed. A successful policy narrative – one that is memorable, taken up and integrated into policy and public discourse – cuts through complexity and heterogeneity, and sets nuance aside. In this way it provides a compelling and powerful framing, a justification and call to arms. It is particularly important to note that a successful narrative will foreground certain solutions or interventions (or development pathways), while explicitly or implicitly delegitimizing others.

Narratives provide a lens through which to view and make sense of a complex and perhaps threatening problem. Successful, compelling narratives are often constructed around a memorable word or phrase: for example, phrases like 'youth bulge', 'demographic dividend', 'farming as a business', 'digital native' and 'waithood' are at the core of the key narratives about rural youth in SSA. Narratives are about communication and persuasion, and they are acutely political. They are constructed, disseminated and used with the aim of promoting a particular perspective on a problem and a set of preferred solutions. As such, a narrative will serve or advance the interests of some individuals, groups and coalitions, while seeking to thwart the interests of other actors.

Policy narratives can be thought of as dominant (hegemonic) or alternative (emergent). However, it is usually not useful to think of them as true or false, right or wrong. Around all important development issues – like rural youth in SSA – there is just too much heterogeneity, too many unknowns and too many legitimate differences in perspective, for any necessarily simplistic narrative to be true in all or most contexts. Ultimately, this does not matter because the job of a narrative is not to convey truth, but to be believable, to stimulate and facilitate a policy response, and to promote certain responses over others. It is nevertheless important to critically examine policy narratives with the aim of understanding, for example, how they foreground or background different groups (e.g. male or female youth) in a variety of rural situations (e.g. high or low potential areas), and how they drive policy responses in particular directions (e.g. toward the youth themselves and away from structural problems). How narratives are used to advance the interests of some groups over others is a particularly important area for research.

Specifically, this book, with its focus on youth in the rural economy, is interested in (i) how dominant narratives align with the different realities of young people's lives in a range of rural contexts; (ii) how they promote certain possible responses and close down discussion of others; and (iii) the politics around their use. This approach to development narratives is different from fact checking, 'myth busting' or 'telling myth from fact' (Christiaensen, 2017; Christiaensen and Demery, 2018; Mabiso and Benfica, 2019). While these exercises are also important, they often fail to appreciate the political nature of policy narratives, and that in policy processes, 'a good narrative is worth a thousand facts'.

The relationship between narrative and evidence is complex and often awkward: an evidence-based narrative is not necessarily the most desirable or the most powerful tool. Too much attention to the detail and nuance of the evidence, the sense that every individual story or village is unique, makes it impossible to construct a strong narrative. This is why 'essentialism' is at the core of the most powerful policy narratives. Phillips (2010, p.47) defines essentialism as 'the attribution of certain characteristics to everyone subsumed within a particular category'. In the narratives addressed in this book, essentialism is expressed through statements like 'African youth are…', 'rural areas in SSA are…', 'agriculture in SSA is…' and 'Africa's youth bulge is…'. Essentialism is de rigueur for a compelling policy narrative, but it provides a very poor basis for evidence generation, policy development or investment decisions.

As will become apparent, and despite the recent upsurge in published work, there is little direct evidence with which to cleanly interrogate some of the most important narratives around youth and the rural economy. The challenge is magnified by a lack of clarity around key concepts and categories (i.e. 'youth', 'migration' and 'aspirations'), and the considerable heterogeneity both among young people and rural spaces. A closely related challenge is that because the evidence base is so patchy, research findings from a detailed study in a particular setting can subsequently be projected across an entire region, country or the whole subcontinent. While nationally representative household survey data address some concerns (see Chapter 2, this volume), they also raise others (Carletto and Gourlay, 2019).

Key narratives about rural youth in sub-Saharan Africa

Debate about, and actions to address, the challenges associated with youth in rural SSA are framed by a number of powerful and persistent, and in some cases, contradictory narratives. This section introduces seven of these narratives that are central to this book and that are taken up in more detail in subsequent chapters. To a greater or lesser extent, they are linked together, and in some cases, they overlap: in both public and policy discourse they are often combined.

Box 1.1. Africa's 'youth bulge' – a defining challenge of our time.

What is the problem? SSA is experiencing a historically unprecedented 'youth bulge' (a very high proportion of the total population being within a specified age bracket, such as 15–25). The subcontinent's resulting youthfulness is associated with both opportunities (the potential 'demographic dividend') and threats (e.g. un- or underemployment, increased international migration, risks of civil unrest and radicalization). A large population of disaffected African youth could have significant negative domestic and international repercussions.

Why or how has it arisen? A slow and late demographic transition.

How should the problem be addressed? Given that the majority of young people in SSA still live in rural areas, agricultural and rural policy will be particularly important if policy makers are to capture the opportunities and avoid the threats associated with the youth bulge. Specifically, they must invest in rural areas, invest in rural young people and promote agroindustry.

This is the central narrative that frames every aspect of the current discussion about African youth. It is particularly compelling because it portrays a potentially dangerous, 'on-rushing future' (de Wilde, 2000; Jansen and Gupta, 2009). This view is premised on a conceptualization of youth, and in particular, unemployed male youth, as rebellious and a threat to domestic social and political stability, and to international relations through uncontrolled migration. Female youth are rarely captured in this narrative, except if they are seen to transgress sexual and moral boundaries voluntarily or through coercion. The link between the youth bulge, youth unemployment and security has been part of the academic narrative for almost two decades (cf. Cole, 2011) and was also highlighted in a speech by Ghana's President, John Mahama, in 2013:

> We need to take the issue of youth unemployment very seriously, so every country should put youth unemployment on its national security agenda. Because if plans are not rolled out to ensure that you engage the youth then you can have a problem in terms of destabilisation and social deviancy.[1]

However, within the narrative, the threat is neatly offset by the potential for a 'beckoning future' that is prosperous and peaceful. For the beckoning future to become a reality, the 'demographic dividend', a one-off economic windfall associated with the youth bulge generation successfully entering the labour market or becoming entrepreneurs, must be realized (Drummond *et al.*, 2014).

Debates around this narrative address both the threat and the promise. There is, for example, disagreement about the potential size and uniqueness of Africa's youth bulge (Bloom and Williamson, 1998; Yazbeck *et al.*, 2015; AfDB, 2016; Baah-Boateng, 2016). There is also considerable contestation regarding the purported relationship between youth unemployment, civil unrest and radicalization (Brück *et al.*, 2016), as well as the potential magnitude of and likelihood of achieving the demographic dividend (Eastwood and Lipton, 2011; UNFPA, 2014; Yazbeck *et al.*, 2015; Ahmed *et al.*, 2016; Losch, 2016; Bloom *et al.*, 2017).

Box 1.2. Youth are leaving rural areas en masse.

What is the problem? Large numbers of, particularly male, youth are leaving their home rural areas and migrating to towns and urban centres. This poses a threat to the agricultural sector and food security, to rural communities, to the migrants themselves who are vulnerable in their new urban surroundings, to urban areas, and to political stability.

Why or how has it arisen? Long-term neglect of rural areas (urban bias) has left these areas devoid of infrastructure and services (water, electricity, health, communications). School curricula neglect (or worse, denigrate) farming and rural life. All things urban are glorified in the media. There is a lack of successful rural role models.

How should the problem be addressed? By making rural areas more attractive through investment in infrastructure and services; by supporting agricultural modernization and agroindustrial development; by changing young people's perception of rural areas and agriculture (i.e. 'mindset change' and sensitization); by better equipping young people to take advantage of the abundant rural opportunities (i.e. train them and build their skills).

At the heart of this narrative is dissatisfaction, and the idea that it breeds within the yawning gap between young people's rising aspirations, and their perception of the limited opportunities available to them in rural areas. Specifically, because of increased educational opportunities and digital connectivity, too many rural young people have had their eyes diverted toward post-secondary education, professional jobs and urban life. While migration of young men is sometimes acknowledged as a 'rite of passage' – part of becoming an adult – and remittances can be invested in the rural economy, overwhelmingly, it is the negative effects of migration that are emphasized. This is a straightforward crisis narrative, with migration portrayed as a threat to everything from the agricultural sector to the young people themselves. It is also a narrative that is manifestly gender blind, referring to youth as a gender-neutral category but representing only the male experience. The female experience of migration or of leaving rural areas, within the framework of marriage or the extended family, is seldom mentioned.

There is much to be considered in this narrative. Migration and mobility – in all their forms – have been well-established facts of African rural life for many decades. Young people leave home for many reasons, including to access schooling and a broader range of educational opportunities. In many parts of rural West Africa, for example, short distance, seasonal movement has long been central to young people's efforts to build their livelihoods. Historically, these mobilities are gendered; young men often begin their migratory trajectories by working on farms and in mines, while young women mostly take up domestic work in urban areas, first for a relative then moving into other waged work as they gain skills (Jacquemin, 2012; Lesclingand and Hertrich, 2017). Whether their absence affects farming depends on the gender division of labour on the farm and on collective and individual inclinations to facilitate a return to work on the family farm during the labour-intensive periods (Linares, 2003). However, their remittances are important factors in some families' relocation to rural towns and their reliance on hired farm workers or sharecroppers.

Equally problematic is the lack of direct evidence that the rate of youth migration has increased (indirect evidence on changing migration rates is provided by: FAO, 2015; Jedwab *et al.*, 2017; Arslan *et al.*, 2018), or that there has been significant change in types or forms of migration. Similarly, there is no evidence of widespread rural depopulation. In any case, not all dissatisfied youth are able to leave, even if they want to, because they lack the social networks or financial resources. Finally, a significant proportion of rural migrants go to other rural areas (Mberu, 2005; Potts, 2013), suggesting that the aversion to agriculture and rural life is overstated.

As will become evident in the chapters that follow, there are many young people actively building livelihoods in rural areas, and they do not universally or generally express a wish to leave.

Box 1.3. Youth do not want to farm.

What is the problem? Young Africans are turning their backs on farming. This is a problem for the agricultural sector and food security; and for the young people themselves, because for some decades to come, only agriculture and agrifood industries will be able to provide the employment opportunities they so badly need.

Why or how has it arisen? Failure of smallholder agriculture to modernise, to embrace technology, mechanization and markets; as a result, farming remains hard, dirty and poorly paid work. School curricula neglect (or worse, denigrate) farming and agriculture. All things urban are fetishized in the media. There is a lack of successful rural role models.

How should the problem be addressed? Use policy to make agriculture economically attractive; change mindsets so that farming is approached 'as a business'; promote engagement with value chains; promote the use of technology (agricultural and digital); reduce drudgery; provide training and develop new skills; make farming 'sexy'.

This narrative is closely linked to the previous one, which suggests that large numbers of young people are leaving rural areas. It also highlights the gap between rising aspirations and the realities of much smallholder farming: hard, dirty, physical work, with poor and uncertain returns, and no respect or recognition from the broader (read 'urban') society. A more nuanced version of the narrative suggests that the problem is not with farming per se, but rather that young people do not want to farm *like their parents*.

There are a number of studies suggesting that a significant proportion of rural youth do not aspire to become farmers (Anyidoho *et al.*, 2012; Petesch and Rodríguez Caillava, 2012; Tadele and Gella, 2012; Leavy and Hossain, 2014; Temudo and Abrantes, 2015; BMZ, 2017; OECD, 2017; Yeboah *et al.*, 2017; Elias *et al.*, 2018). Debates revolve around different understandings of aspirations, and the degree to which aspirations drive (or even inform) choices, decisions and/or outcomes. However, there is also some literature suggesting that in some situations young people are actively pursuing or have an interest in agricultural livelihoods (e.g. Berckmoes and White, 2014; Andersson Djurfeldt *et al.*, 2019; Ruiz Salvago *et al.*, 2019; Yeboah *et al.*, 2020).

Box 1.4. Youth want to farm but cannot access land.

What is the problem? Young people are leaving agriculture and rural areas, not because they want to, but because they cannot access land. In effect, they are 'land scarcity migrants'.

Why or how has it arisen? Increasing rural population density creates pressure on land; traditional inheritance rules result in fragmentation of holdings; processes of commodification block channels through which young people have traditionally accessed land; the older generation will not make unused land available to the younger generation; and land markets are generally weak or non-existent.

How should the problem be addressed? Provide young people with privileged access to land to get them started; support development of land markets; change inheritance rules.

This narrative runs counter to elements of the narratives 'youth are leaving rural areas' and 'youth do not want to farm' outlined above. Critically, it stresses 'push' factors (limited access to productive resources, and thus economic opportunity) as opposed to 'pull' factors (for example, the lure of professional jobs and urban environments). Rooted in historical, evolutionary and political economy perspectives, the research that underpins this narrative is not preoccupied with changing aspirations, but rather with the changing agrarian context. The argument is that increases in population density, processes of commodification (of crops and land) and associated changes in generational relations are making it increasingly difficult for young people to access land and begin to build an agrarian livelihood (Amanor, 2010; Berckmoes and White, 2014; Berckmoes and White, 2016; Kosec *et al.*, 2016, 2018; Scoones *et al.*, 2019).

An extension of this narrative suggests that even if land is available, an inability to access credit, technology and markets inhibits young people's farming. The implication here is that young people are being discriminated against simply because they are young. However, the argument can be made that in many rural areas, nobody – young or old – has much access to formal credit, etc., and thus the fact that young people do not have access is not a very compelling basis for policy or intervention.

Box 1.5. Rural areas in sub-Saharan Africa are brimming with opportunity that young people just do not see.

What is the problem? Growth in international, regional and national demand, linked to urbanization and changing patterns of consumption, creates significant opportunities for producers and processors of agricultural products in Africa. However, rural young people are unaware of these many and varied opportunities for work and livelihood building. This is a problem because it fuels dissatisfaction and rural out-migration, and robs the rural economy of the 'best and brightest'.

Why or how has it arisen? Historic neglect of rural areas; school curricula that denigrate farming and rural life.

How should the problem be addressed? Develop 'inclusive' agricultural value chains; build an entrepreneurial culture; revise school curriculum to highlight rural opportunity and role models; raise awareness; and invest in training and skill development.

This narrative, which tends toward a new 'rural prosperity gospel' (Sumberg *et al.*, 2020) is rooted in a vision of a modernized, transformed agricultural sector. The vision bundles together investment in infrastructure, with the use of new technology (from seeds and mechanization, to digitally delivered weather and prices), entrepreneurship, engagement with agricultural value chains, and perhaps most important of all, the emergence of a new ethos of 'farming as a business'.

However, in many if not most rural areas there is a real gap between this future vision and the current reality. This then raises the question: 'Even if the investment and political commitment are forthcoming, when will future generations of rural youth actually benefit?' Acknowledgement of the gap between vision and reality leads to a somewhat less optimistic variant of the narrative: for some years to come young people will find work in and around agriculture because they will continue to live in rural areas, and there is nothing else that can provide employment on the scale needed (Filmer and Fox, 2014). In this view, engagement with the rural economy is about making the best of a less than optimal situation; in many ways, as will become apparent in the chapters that follow, this sums up much of what is observed across the subcontinent.

Box 1.6. Youth hold the key to rural transformation.

What is the problem? Rural economies in SSA are already going through a transformation that includes a shift from production for own consumption to production for the market, increasing productivity, value upgrading and the like. However, the pace of transformation is very slow, which results in continuing poverty, dissatisfaction, pressure to migrate, environmental degradation, etc.

Why or how has it arisen? Technology promotion efforts have been focused on older, principally male, farmers (the 'household head'), but these individuals are often conservative, have little motivation to innovate, and just do not understand the opportunities associated with 'digital agriculture'.

How should the problem be addressed? Target technology promotion efforts at young people because they are innovative and quick to adopt new technology; also, because they are 'digital natives', young people are particularly well placed to turn the promise of digital agriculture into reality.

This narrative is constructed around the essentialist proposition that youth – as a group – are different (i.e. from older people), in that they are particularly innovative and have a special orientation toward the adoption of technology (cf. Sumberg and Hunt, 2019). Given the widely held assumption that technological change will drive productivity enhancement and rural transformation, it is then only logical that young people must be at centre stage. Indeed, an argument that is associated with this narrative is that without a special focus on young people and their capacity to act as positive 'agents of change', there will likely be no transformation.

In addition to linking to broader debates about the relationship between age and innovative or creative behaviour, this narrative suggests that it should be possible to observe meaningful differences in how young people farm in SSA compared to older people (see Chapter 6, this volume).

Box 1.7. Youth are stuck in 'waithood'.

What is the problem? The majority of young Africans are living in 'waithood', 'a prolonged and uncertain stage between childhood and adulthood that is characterized by their inability to enter the labour market and attain the social markers of adulthood' (Honwana, 2012, p.19).

Why or how has it arisen? The social contract between the state and its citizens has broken down because of unsound economic policies, bad governance, corruption and the erosion of civil liberties. This breakdown prevents young people from transitioning to adulthood and becoming active, fully-fledged citizens.

How should the problem be addressed? Young people are in waithood but they are not passive. Through their involvement in social and political movements they are rejecting formal politics and the corruption that characterizes it; protesting and making revolution; challenging the modernity project; and negotiating new terms of membership into the global community.

In the book *The Time of Youth: Work, Social Change and Politics in Africa* (2012), Alcinda Honwana used the notion of 'waithood' to capture a sense of disrupted and delayed youth transitions ('waithood' had been used earlier by Singerman (2007), and Dhillon and Yousef (2009)). Based primarily on interviews conducted in Mozambique, Senegal, South Africa and Tunisia, she described waithood as:

> [...] a prolonged and uncertain stage between childhood and adulthood that is characterized by their inability to enter the labour market and attain the social markers of adulthood [p.19]. [...] Waithood is a neither-here-nor-there position in which young people are expected to be independent from their parents but are not yet recognized as social adults. No longer a brief transitional stage in the life-course, waithood is becoming a permanent condition, as many

> young people remain stuck in this in-between situation. Indeed, waithood is becoming a new but socially attenuated form of adulthood [p.20].
> (Honwana, 2012, pp.19–20)

Along similar lines, but based on work in India, Jeffrey (2010) talks about young people 'doing timepass' ('a means through which young men mark their social suffering and begin to negotiate unemployment' (p.477)). And Sommers (2012) suggests that young people in Rwanda are 'stuck' due to a lack of jobs and restricted access to productive resources that draw out or postpone transitions from youth to adulthood.

Honwana made and has repeated a specific claim – that 'the majority of young Africans today live in waithood' (Honwana, 2012, p.20; Honwana, 2019, p.8). This is a very big claim: taken at face value it suggests that across Africa there are at least 210 million people between the ages of 15 and 35 who are living in waithood (UNEP, 2019), and unable to move on with their lives (although, as Honwana stresses, they are not inactive). The claim is important because it provides the foundation for her arguments and reflections concerning young Africans' aspirations, economic activities and citizenship, and the steps that should be taken to address Africa's youth crisis (see Chapter 8, this volume). It is also significant because it has become integrated, as fact, into policy discourse.

The Argument

As these narratives are used to justify and promote particular interventions (and marginalize others), they have a profound effect on policy processes and the choice and design of interventions. This book is organized around the most prominent of these narratives.

The argument developed through the chapters that follow is that:

- As should be expected, given the heterogeneity in people and place, there is a significant disjuncture between the dominant narratives around young people's engagement with the rural economy, and the diversity of what is observed across rural SSA.
- Specifically, the central story is not about young people leaving rural areas and the rural economy, but rather about the many millions who are working hard, often in difficult conditions, to build rural-based and rural-inclusive livelihoods. There are few signs that these young people are stuck in permanent waithood, or generally looking to leave rural areas.
- These efforts, and the resulting livelihoods, are: gendered in important ways; reflect widespread disappointment in and failure of the education system; often combine on-farm, off-farm and non-farm activities; and in many cases the agricultural engagement does not appear to be prioritizing 'farming as a business' and/or engagement with agricultural value chains.
- They also reflect the central role of hazard, events or idiosyncratic shocks in the lives of rural youth (as distinct from the shared shocks that, for example, typify rainfed smallholder agriculture). This raises important questions in relation to thinking about aspirations, preferences, choice, decision making and the exercise of agency. It also poses significant challenges for development policy and practice.
- In contrast to the dominant narratives, most rural young people imagine futures for themselves in which agriculture and/or the rural economy play important roles. However, in these imagined futures they are often not actually doing the farming, but rather managing labour while they themselves pursue trading or other activities.
- Finally, we argue that these insights have profound implications for youth policy, social policy, agricultural and rural policy. Specifically, there is an important disjuncture between the vision of the rural economy and rural livelihoods that underpin policy, and the futures that young people imagine and are in the process of enacting for themselves and their communities. Rethinking the current tendency for constructing rural youth as another 'target group' for development, to be served by youth-specific interventions, should be a top priority.

Conceptual Grounding

This section provides a brief introduction to the concepts and frames that underpin the book's analysis of young people's engagement with the rural economy. These are further developed and nuanced in the individual chapters.

Generational and life course perspectives on youth[ii]

A generational perspective offers a way to understand youth as belonging to a distinct social group, and in relation to other social groups (Hopkins and Pain, 2007; White, 2015; Huijsmans, 2016). Generations were established as a subject of interest by Karl Mannheim, when he articulated the notion that a cohort of people born around a similar time may share certain formative experiences, perspectives, relationships and identities that situate them uniquely in relation to other generations (Mannheim, 1952). Everyone is part of a generation and is therefore enmeshed in intra- and intergenerational relations. Cultural norms, as well as negotiations, struggles and outright conflict between generations, define the mutual rights and obligations of each generation in relation to others – an 'intergenerational contract' (Huijsmans, 2016). In particular, relationships with older generations strongly determine whether, when and under what terms, a member of a younger generation can obtain and use resources, express her or himself, exercise independent choices and make decisions, and build a different kind of life. The concept of 'social navigation' has been proposed as a way to examine how young people attempt to negotiate intergenerational relations, to 'disentangle themselves from confining structures, plot their escape and move towards better positions' (see also Christiansen *et al.*, 2006; Vigh, 2006, 2009, p.419).

In addition to being part of a generation, each individual life passes through a succession of phases, each being both distinct (i.e. marked by specific experiences and challenges), and a time of evolution and change that connects the preceding and following phases. Life phases and the transitions between them are socially and culturally constructed. Few real lives unfold precisely in the ways anticipated by bureaucratic, medical or legal categories (such as child, adolescent, young adult; above or below the age of majority, etc.), which are usually based on age-defined norms and assumptions. The youth phase is typically framed as a period of particularly rapid and fundamental transition, characterized by physical and cognitive growth and transformation, a great deal of learning, a substantial expansion of social networks and the building of social capital. In many if not all societies, marriage and child-rearing are normal expectations of youth, or key signifiers of a transition between youth and adulthood. For most individuals, the transition from childhood through youth to adulthood is marked by increased independence, autonomy and responsibility. These transitions are usually relative and incremental rather than sudden and complete, yet in some contexts, such as rural SSA, many young people and especially young women, ironically, do not experience this period of 'youth' or only do so to a limited extent. Rather, they transition more directly from childhood into something more akin to adulthood (with childbearing and marriage sometimes coming during adolescence, teenagers leaving school and becoming economically active and/or independent, forming independent households, etc.).

While there may be important legal, bureaucratic and policy reasons for defining youth as sitting between specific lower and upper age boundaries,[iii] this makes little sense from a sociological or cultural perspective (see, e.g. Bourdieu, 1993). Our focus in this book is on young people in the early stages of livelihood building. Practically speaking, this means that of the 416 people who participated in individual interviews in the main International Fund for Agricultural Development (IFAD) Youth Study (see Chapter 2, this volume) 86% were less than 30 years old. For the analysis of nationally representative survey data, where clear age cut-offs are required, youth have been defined as being between 18 and 24 years old.

In line with the generational perspective, in this book we see young people not as isolated, autonomous actors, but as deeply enmeshed in diverse social and economic relationships. The vast majority are first and foremost part of families, and initially largely dependent on others (their parents, carers and kin), and later with families of their own and possible links to the family networks of their partners. The importance of extended family and clan networks for young people in SSA has been widely recognized (Langevang, 2008). These networks can enable and facilitate, as well as constrain, and as with interhousehold processes more generally, they are characterized by both collaboration and conflict (Seymour and Peterman, 2018; Acosta et al., 2019).

Family networks, including but not limited to parents, enable young people to attend school, are primary sites of skill acquisition beyond formal education (Yeboah et al., 2020), and provide access to land and capital (Flynn and Sumberg, 2017). Existing studies have documented the importance of kin networks in migration, and in finding work, securing apprenticeships and in accessing educational opportunities away from home (Langevang, 2008; Thorsen, 2013). Young women and men have expectations placed upon them and acquire various responsibilities toward other family members, like contributing to the costs of siblings' education and taking care of old and sick family members. While living independently in one's own home and being able to provide for the family is central to many young Africans (Sommers, 2012), being able to fulfil such family-oriented expectations is also a priority.

In recognizing the importance of social relationships, it is critical to also acknowledge the social dynamics that create and maintain hierarchies and markers of social difference, both within and outside the household. Gender relations and behavioural norms interact and create dynamics that strongly shape the livelihood activities considered appropriate for young women and men, the distribution of labour and care responsibilities, land access and autonomy over earnings. Gender norms also come into play in decisions over whether and how to support the education or enterprises of daughters and sons, sisters and brothers, and female and male partners, as well as decisions about migration (Chant and Jones, 2005; Carr, 2008; Doss et al., 2015; Elias et al., 2018; Van den Broeck and Kilic, 2019). In addition to shaping all aspects of livelihood building, from aspirations and access to education and productive resources, to opportunities for mobility, these norms play a central role in setting out what it means to be a 'good woman' (Elias et al., 2018) and a 'good man'.

School and education

A key axis of our analysis is the role that education plays in lives, livelihoods and imagined futures of rural youth in SSA. This is of particular interest because education has been viewed as central to ensuring that young people's potential as economic actors and citizens is directed in productive ways. In response to the Millennium Development Goals (MDGs), efforts were made across SSA to develop free universal basic education. However, the rapid expansion of the schooling system in many contexts proved to be at the expense of education quality: primary enrolment in SSA soared by 75% between 1999 and 2012 (UNESCO, 2015) but this increase was not matched by the development of school infrastructure or investment in teacher education. In addition to exclusions related to the quality of provision, location, religion, language of instruction and ethnicity, practices of schooling are deeply gendered, and schools are a persistent site of gender violence (Dunne, 2007; Humphreys et al., 2008). Many of these issues are legacies of the unequal ways that schooling was developed in SSA during colonial times (Dunne and Adzahlie-Mensah, 2015), which mean that education continues to benefit urban more than rural populations, the rich rather than the poor, and males rather than females (UNESCO, 2018).

Human capital theory (HCT) is the dominant lens through which relationships between education and work have been considered. Many critiques of both the evidence base and assumptions of HCT have been voiced (Bennell, 1996; Marginson, 2019; McGrath et al., 2019). In education (and work), multiple forms of capital come into play, which are valued in ways that are context dependent, relational rather than universal, and not reducible to individual deficits (e.g. The World Bank, 2014). Attempts to isolate 'causal' factors in an overgeneralized way fail to attend to their relational intersections, and indeed the relational character of all social life. As Bourdieu and Passeron (1990) argue, it is 'the system of factors, acting as a system' which requires analysis; for them it is 'absurd to try to isolate the influence of any one factor' (p.87).

Alongside the complex intersections of the fields of education and work, Bourdieu's theoretical frameworks generally offer important counter arguments to the individualization of deficit in dominant policy narratives about youth and education (see Chapter 7, this volume). Bourdieu illuminates the power of schooling to produce and arbitrate on what comes to be seen as 'individual merit', dressed in the language of ability, skills and competences. His theories show how education reproduces dominant norms and rationalities and gives legitimacy to social stratification. He also helps understand the shifting value of educational qualifications, whereby as more young people gain qualifications, their value in the job market diminishes, making social capital and the inherited cultural and economic capital of the family additionally significant in the struggle for positioning in the market (Brown *et al.*, 2011). Yeboah *et al.* (2017) and Ansell (2018) signal how the imaginaries of youth in different contexts in the Global South often hold an idealized notion of professional work which merits further exploration. Overall, however, there remains a paucity of knowledge about rural youth's perspectives on their education and how these relate to their imagined futures.

Mobility

The current moral panic in Europe, the US and elsewhere around international migration reflects a similar, long-standing and equally ambivalent stance toward the migration of rural people in SSA. The historical sacrifices and contributions of migration and migrants are celebrated, while today's migrants and would-be migrants are demonized. Further, with specific reference to SSA, Kleist and Thorsen (2017) describe a 'mobility paradox' that arises because increased access to transportation and communication infrastructure exposes young people to seemingly attractive livelihood opportunities elsewhere, at the same time as global inequality and restrictive migration policies exclude (or discourage) many from participating even in legal circuits of mobility.

The orthodox view frames rural young people as either 'international' or 'internal' migrants based on whether they move within or beyond their national borders, or as 'left behind' if they remain in their community. Numerous other categories are commonly used to describe migration and migrants in SSA: independent, first- or second-generation, permanent, temporary, seasonal, rural-to-urban, rural-to-rural, stepwise, return, voluntary, forced, and many others. According to Van Geel and Mazzucato (2018), placing young people in one of these categories, as is done in many studies, obscures the variation in mobility that characterizes young people's biographies, and this variation is central to understanding the ways in which mobility is used by, and impacts the lives of, rural youth. As is increasingly the case in the academic literature, in order to escape the intellectual tyranny of these categories and the discourse they help to frame, the focus in this book is on mobility as opposed to migration. Whether as movement in search of land for agricultural production, cyclical livelihood mobility, movement to towns in search of employment in the formal or informal sectors or entrepreneurial opportunities, or daily journeys to work (e.g. fishing), to school, or to meet friends, relations and business operators within the rural economy, mobility is – and has long been – as much a part of rural life in SSA as farming. It is central to the lives of many rural youth in SSA and shapes their experiences and future life chances.

Migration demands mobility, but mobility that is part of livelihood building, or that enables beneficial engagement with the rural economy, does not necessarily mean migration.[iv] The literature on mobilities speaks of 'achieved mobility' as a source of excitement, inclusion, thrills, temptation and perceived success but also as a cause of fear, danger and exhaustion. On the other hand, 'frustrated mobility' is regarded as a source of exclusion, despair, anger, and perceived failure (Porter *et al.*, 2010). While the motivations driving mobility, and the associated outcomes, involve complex, multiple and diverse realities, the anthropological and social geography literatures provide a generally positive view of mobility as a means for young people to access social and economic opportunities, as a rite of passage, a way to strengthen kinship ties, search for autonomy, and facilitate the transition to adulthood (Hashim, 2005; Hashim and Thorsen, 2011; Beazley, 2015; Van Blerk, 2016).

The authors of *Youth and the Rural Economy in Africa: Hard Work and Hazard* do not take a normative moral stance, explicit or implicit, that rural young people are or would be better off if they stayed in the countryside, or that reduced rural-to-urban mobility is necessarily good for society, the economy, urban areas or rural communities.

Imagined futures, future selves, aspirations

The aspirations of young rural Africans have received some attention in the rural development research literature, motivated by the understanding that: (i) increasing levels of education and connectivity fuel rising aspirations; and (ii) to a greater or lesser degree, aspirations inform decisions about education, employment, marriage, migration and so on.

This literature generally leaves unquestioned the conceptual and theoretical dimensions of aspirations and their links to behaviour, and the associated methodological and interpretive challenges. In contrast, such dimensions are important concerns within the sociology and youth studies literatures (Zipin *et al.*, 2015). Hardgrove *et al.* (2015), for example, argue that alternatives to aspirations, including 'imagined futures', 'future selves' and 'possible selves', deserve attention because they provide 'a theorization of the link between imagined possibilities in the future and motivation to act in the present' (p.163). Specifically, they argue that the notion of possible selves provides a pathway to investigate how imagined futures become motivational, influencing the actions of young people, not only in the immediate but also toward the imagined. This thus provides a broader theoretical understanding of the agency of young people as they negotiate (navigate) transitions within the present and toward the future. The idea of possible selves is that they are underpinned by individual and social experiences, and combine a mixture of conceptual explanations that dwell on the value of opportunity structures, motivations, socio-cultural based meanings and self-concept (Erikson, 2007). Thus, while aspirations as a concept may appear 'vague' and 'inflated', and not very well grounded in structured positions within society, possible selves are more rooted in present circumstances and past experiences. They 'encompass within their scope visions of desired and undesired end states' (Markus and Nurius, 1986, p.959), and are rooted in personal exposure and experiences, as well as interpersonal relationships that could enable the individual young person to imagine what her or his life could become.

Key to the debate about young people's engagement with the rural economy is the extent to which imagined futures or aspirations shape the livelihood building process. In contrast to much of the literature on youth in SSA, the analysis in this book privileges imagined futures over aspirations, as a window on young people's thinking about their futures in general, and their engagement with the rural economy in particular.

Opportunity structures, agency, hazard and performance

Mainstream microeconomic theory is built on the notion of a rational, utility-maximizing individual who, within some bounds, is free and able to exercise choice. Information availability, the nature of the decision making processes, and how to avoid 'bad' decisions, are key concerns. This rational choice perspective is challenged by sociological perspectives that emphasize the embeddedness of individuals in social relationships. Additionally, within applied economics and political science work in developing country contexts, an increasing prevalence of experimental work informed by behavioural economics has identified many areas where human behaviour deviates systematically from that predicted by standard neoclassical economic theory (Carter, 2016; Kremer *et al.*, 2019; Streletskaya *et al.*, 2020).

In relation to the initial phases of livelihood building, a useful starting point is the theory of 'occupational allocation', also referred to as opportunity structure theory (Roberts, 1968, 2009). The central tenet is that the job opportunities available to school leavers become 'cumulatively structured', and Roberts theorized that what he called 'opportunity structures' act to create distinct routes that govern both young people's entry into the labour force and subsequent career progress. These opportunity structures emerge from a web of determinants including place, family origins, gender, ethnicity and education, and labour market processes. It is not so much that opportunity structures leave the individual with no room for manoeuvre, but rather that for most young people who are poor, poorly educated, and/or socially or geographically marginalized, it is likely that their room for manoeuvre will be very tightly constrained. Due to this structuring, Roberts (1977) argued that it is a mistake to overemphasize the role of aspirations and choice in determining how young people enter the labour market, particularly in landscapes of constrained opportunity. Indeed, he put it even more starkly: 'Neither school leavers nor adults typically choose their jobs in any meaningful sense: they simply take what is available' (p.3). Some years later he elaborated:

> Choice is not irrelevant, but it fails to explain enough. It cannot account for the contexts, including the labour market contexts, in which young people make their choices, and it cannot identify the different limits within which different groups of young people choose.
> (Roberts, 2009, p.362)

The main implication of opportunity structure theory is that aspirations, choice and individual responsibility are simply not very useful foci for policy. Change in how young people enter and progress in the labour market will come about, not as a result of higher aspirations, altered preferences, better choices or better skills; rather, it is the opportunity structures that need to shift, and this implies fundamental social and economic change.

Bourdieu's concepts of 'habitus' and 'field' are particularly relevant as they capture the relationship between individual agency and social forces. *Field* encompasses the structures, standards and norms in the environment, and *habitus* is the social process through which individuals become socialized (Bourdieu, 1984; Navarro, 2006). Habitus is neither a result of free will, nor overly determined by structures, but rather created by the interplay between the two, over time (Bourdieu, 1984, p.170). Habitus is created and reproduced unconsciously, 'without any deliberate pursuit of coherence... without any conscious concentration' (Bourdieu, 1984, p.170). Moreover, habitus is about a 'full internalization of social experience' (Pettit, 2016, p.96): past experiences, beliefs and norms are embodied, and influence people's ideas and practices. Therefore, while a rational choice perspective assumes that individuals will consider, even experiment with, all possible actions, through habitus the individual gives disproportionate weight to (or perhaps only sees) some of these options, influenced by past experience that has shaped her or his very rationality (Bourdieu, 1980; Pettit, 2016). Habitus thus influences perceptions and preferences and here we can link back to the point made above: gender norms influence young women's and young men's aspirations and imagined futures.

In addition to ideas around the interplay of structure and agency, we also draw on the notion of 'hazard' as developed by Richards (1986). Richards conceived of hazard as including accidents, weather events and mistakes that have (or could have) negative impacts, and in some cases, these might be cumulative. For example, a family illness that then affects farming operations represents hazard, as would early rains that disrupt land preparation, or a decision to plant a late maturing crop variety in what turns out to be a dry year. The notion of hazard has much in common with the idea of 'idiosyncratic risks' (or idiosyncratic shocks) in the economics literature (Dercon and Krishnan, 2000). While common or covariate shocks affect whole communities or countries by reducing household income, consumption, and/or the accumulation of productive assets, an idiosyncratic shock affects only one individual, household or family. Idiosyncratic shocks can arise in the economic sphere (e.g. crop failure or loss of employment), or through ill health, family break-up and crime. Dercon (2002) notes that 'even within well-defined rural communities, few risks are purely common or idiosyncratic' (p.143). An important difference shown by Richards between hazard and idiosyncratic shock is that he included mistakes in the former, while they play no part on the latter.

As will become evident, the lives of many rural children, and their subsequent efforts as young people to build their livelihoods, are affected by hazard. The death of a parent or guardian, parental divorce, frequent relocations that disrupt school attendance, a parent's inability or refusal to pay school fees, early pregnancy: these everyday yet potentially life-changing events figure prominently in the lives and life histories of rural youth. For the vast majority, neither insurance nor social assistance programmes, two formal mechanisms for risk management, are currently available.

Closely linked to the hazards of everyday life is Richard's notion of performance (Richards, 1989; Flachs and Richards, 2018). This entails serendipitous discoveries and processes that are the outcome of practices – defined as socially-situated actions – with which people respond to unpredictable events.

Rural economic geography and engagement with the rural economy

Rural economic opportunity exists both on- and (increasingly) off-farm, and it has a strong spatial dimension (Chamberlin and Jayne, 2013). A simple framework for thinking about the diversity of rural areas was proposed by Wiggins and Proctor (2001) who use differences in quality of natural resources and access to markets to characterize current activities and potential future agricultural and rural development trajectories (Table 1.1). In a similar vein, the 'development domains' literature uses agricultural potential, access to markets and population density to understand 'opportunities and constraints facing alternative rural livelihood options' (Pender *et al.*, 2004, 2006; Chamberlin *et al.*, 2006).

Frameworks like these are useful because they make explicit the critically important spatial dimensions of processes of agrarian change including agricultural intensification, commercialization and rural transformation more broadly. In so doing, they provide an important window on to the different 'landscapes of rural opportunity' that confront young people and others (Sumberg *et al.*, 2019; Abay *et al.*, 2020).

Specifically, Table 1.1 leads to the generation of hypotheses regarding the relationship between the level of agricultural commercialization on the one hand, and the landscape of rural opportunity on the other. For the sake of argument, we can consider two ends of a continuum of rural economic dynamism – 'hot spots' (darker shading in Table 1.1) and 'cold spots' (no shading). The stylization shown in Table 1.2 suggests a more diverse landscape of opportunity in hot spots, including high value crops, farm wage work, non-farm wage work and non-farm self-employment. At the same time, it is likely that there will be greater pressure on land in hot spots, which may restrict access of some individuals. It is expected that market-based mechanisms for land access (like rental markets) will be more important in hot spots than in cold spots.

The central concern of this book is with young people's engagement with the rural economy, and the focus is on the work or production side (as opposed to the consumption side) of engagement. Building on the previous sections, Fig. 1.1 shows how this engagement is conceptualized. At the centre of the figure is a local rural economy with some combination of farm and non-farm economic activities. Individuals who farm, gain access to land either through non-market

Table 1.1. Rural diversity: a characterization, with most likely activities. Adapted from Wiggins and Proctor, 2001. 'Accessible' areas include peri-urban and rural areas with good physical access to urban markets.

Quality of natural resources	Location characteristics		
	Accessible areas	'Middle' countryside	Remote rural areas
Good	*Market gardening and dairying* Daily commuting to the city Weekend recreation activities Manufacturing industry may 'deconcentrate' from city proper into this space	*Arable farming and livestock production, specialized, with capital investment, producing surpluses for the market* [Same for forestry, fishing, mining, quarrying] Tourism and recreation Some crafts Employment in off-farm economy Migration (in or out)	*Subsistence farming, with only the production of surpluses of high value items that can bear transport costs* Crafts and services for local markets Tourism and recreation Migration (out)
Poor	As above: i.e. *Market gardening and dairying* NB: Quality of natural resources not so important since capital can be used to augment poor land – e.g. by irrigation, fertilizer, greenhouses – when needed for intensive farming	*Extensive farming, probably livestock* Probably lightly settled Few jobs Tourism and recreation Some crafts Migration (out)	*Subsistence farming, low productivity, surpluses very small or nil* Lightly settled Crafts and services for local markets Tourism and recreation Migration (out)

Table 1.2. Stylized contrast characterization of rural hot spots and cold spots.

Characteristic	Hot spots	Cold spots
Orientation of farming	More commercial	Less commercial
Distance to major markets	Near	Far
Population density	Higher	Lower
Social services, including schools	Better	Poor
Internet service	Better	Poor
Economic dynamism	Higher	Lower
Opportunities for high value crops	More	Few
Agric. input / service availability	Better	Poor
Opportunities for farm wage work	More	Few
Opportunities for non-farm self-employment	More	Few
Opportunities for non-farm wage work	More	Few
Land availability	More restricted	Less restricted

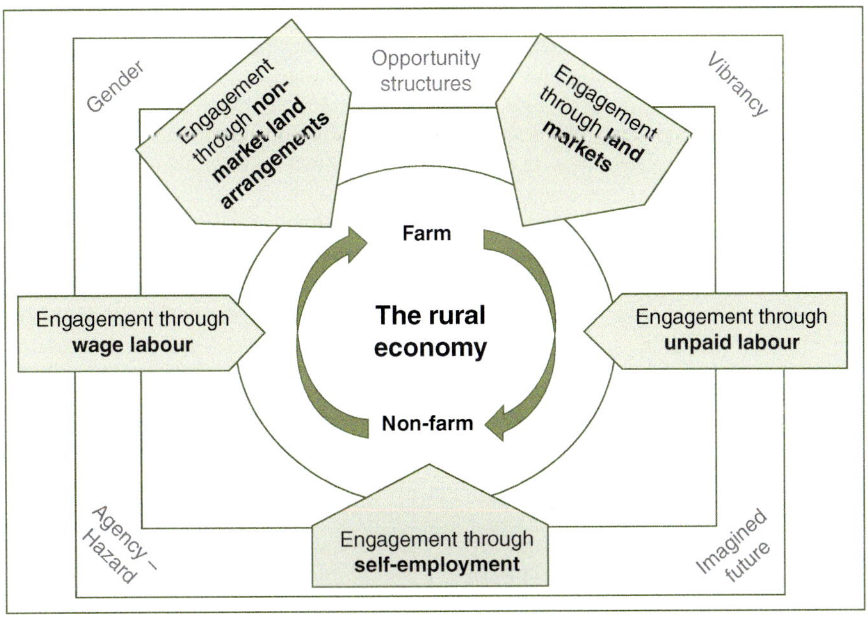

Fig. 1.1. Conceptualization of young people's engagement with the rural economy. Courtesy of the authors.

(traditional land, family land, borrowing, etc.) or market (purchase, rental, sharecropping) arrangements. Three labour arrangements are envisaged in relation to both farm and non-farm activities: unpaid labour, self-employment and wage labour. Finally, the kinds of activities undertaken, means of accessing land, and labour arrangements are mediated by opportunity structures, gender, economic vibrancy, agency, hazard and an individual's imagined future.

Chapter Summaries

Youth and the Rural Economy in Africa: Hard Work and Hazard proceeds as follows.

Chapter 2 (Oosterom, Chamberlin and Sumberg) focuses on the different empirical windows that have been used to study young people's economic lives and livelihoods. A selection of primarily quantitative, primarily qualitative and mixed method studies is analysed in terms of the questions addressed, methods used and the scale of the spatial analysis. Following this, the methods used in the research presented in this book are described and situated in relation to the larger body of literature touching on youth and the rural economy. The final section makes some specific recommendations on how these windows and methods might be improved to gain greater, more policy-relevant insight on young people's lives and livelihoods, in all their diversity.

Chapter 3 (Flynn and Sumberg) is motivated by the oft-repeated claim that rural young people are turning their backs on agriculture. To put this claim into perspective it explores how young women and men engage with the rural economy across a selection of countries and contexts. In addition to unpaid domestic work, farm work and care roles, three broad patterns of engagement (or segments) emerge. Many young people combine some involvement in crop and/or livestock production with off-farm or non-farm employment, while others are not involved in agriculture at all, but engage only in non-farm self-employment and/or wage employment. A relatively small number of others engage only in agricultural activities. The chapter provides examples of these patterns and explores the influence of intensity of agricultural commercialization in shaping them.

Chapter 4 (Chamberlin, Yeboah and Sumberg) argues that land issues play a direct or indirect role in many if not most rural 'youth questions' that are the focus of policy discussion. This includes the changing ability of young people to access land for starting out in farming, driven by increasing scarcity of land, but also includes the broader transformations that are accompanying such scarcity. These transformations include the increasing commodification of land and replacement of traditional access institutions with market mechanisms. They also include expanding rural labour markets and rural mobility, changing farm size and characteristics, and the evolution of agricultural value chains and the farm and non-farm rural economies, all of which are endogenously linked. This chapter lays out what we currently know, identifies areas requiring further empirical research, and some of the data collection innovations that will enable such research.

Chapter 5 (Thorsen and Yeboah) uses the 'mobility paradigm' to shift attention away from the deeply entrenched focus on transnational and irregular migration, and the relatively few rural young people involved in them. Rather, the chapter highlights the central role of movement and relocations over shorter spatial and temporal scales in many young people's livelihoods and transitions. Using interview data, examples of forced movement due to civil unrest, and mobilities for education and/or work are explored, with a particular emphasis on gender differences and the role of social control.

Chapter 6 (Chamberlin and Sumberg) uses household survey data to address three questions: How might we think about the notion that youth bring something new to farming? What aspects of young people's farming are visible with existing empirical windows? Do the young in SSA farm differently? The analysis provides some support for many of the stylized assertions about youth in African agriculture. Young households are associated with marginally higher propensities for engaging with intensification practices and commercial orientations. However, the very limited magnitude of these age effects suggests much caution should be exercised in making the argument that young people's inherent vim and vigour are important and underutilized assets for agricultural growth and transformation in SSA.

Chapter 7 (Crossouard, Dunne and Szyp) focuses on formal education. After laying out the intellectual background for the importance of education in Africa's development agenda, the chapter draws on qualitative research into youth livelihoods in four SSA countries to challenge a number of key assumptions. First, it illuminates the extent to which young people value education. It then turns to their lived experiences of juggling both schooling and work from an early age, highlighting the wide disparity between idealized notions of 'transition' and the complexities of youth livelihoods. It then explores the gendered dimensions of this social landscape, and how these produce different pressures that force young women in particular, out of education. The chapter concludes with implications for young people's current and future engagement with the rural economy, and for education policy.

Chapter 8 (Oosterom) interrogates the increasingly popular notion of waithood, and particularly the idea that most young people are stuck permanently in waithood because they cannot enter the labour market. Based on empirical data gathered from young rural women and men in Uganda, Ethiopia and Nigeria, the meaning of farming and other economic activities in their lives, particularly in relation to social status, is presented. Other avenues for claim making on social recognition, status and respect are then analysed, with a focus on marriage, family life, and active citizenship. Throughout the chapter the gendered nature of the process of becoming a social adult is emphasized.

Chapter 9 (Yeboah, Crossouard and Flynn) explores the futures that rural young people imagine for themselves, and how they relate to both their current engagement with the rural economy, and the narrative that suggests a widely held desire to abandon agriculture and rural areas. Beyond imagined future economic activities, including farm and non-farm work, the chapter highlights the place of education and migration in young people's imagined futures.

Chapter 10 (Sumberg, Szyp, Yeboah, Oosterom, Crossouard and Chamberlin) synthesizes the main findings of the seven empirical chapters, and sets out their discursive and practical implications for policy, research and practice relating to youth, and to agricultural and rural development.

Notes

[i] President John Mahama's presentation at a youth work session in Cotonou, Benin, on Ghana's experience with unemployment and job creation, Friday 13 September 2013, https://www.youtube.com/watch?v=yt4ldpWqxGE, accessed on 16 January 2021.

[ii] This section draws heavily on Glover and Sumberg (2020).

[iii] The UN mainly uses the 15–24 age range – 'for statistical purposes' – as its definition of youth, though some UN entities employ other definitions, such as UN Habitat which uses the 15–32 age range (UNDESA, n.d.). The UN also recognizes that the definition of youth varies across the world, and the International Labour Organization (ILO) itself acknowledges the 'growing momentum' to extend the definition up to the age of 29, given that not all young have completed their education and have thus postponed their entry into the labour market (ILO, 2015, p.1). Finally, the African Union, as stated in its 'Youth Charter', uses the 15–35 age range as its definition of youth (African Union Commission, 2006), though various African governments use different definitions (e.g. Kenya uses the 15–30 age range (Republic of Kenya, 2007), while Malawi uses the 10–35 age range, stating that 'the definition is quite flexible, bearing in mind the variety of parameters that could be used in categorizing the youth' (Republic of Malawi, 2013, p.*v*)).

[iv] For example, Mueller and Lee (2019) use survey data to investigate youth migration, while defining as a migrant anyone who moves out of the household in which they were located in the previous survey round. 'We further compare the distances travelled by migration pattern [...]. The median distance that young (15–24) rural-[to-]rural male (female) migrants travel is 1.4 (1.4) kilometres in Malawi, and 0.2 (1.6) kilometres in Tanzania' (p.31). In other words, more than half of the male, rural-to-rural 'migrants' in Malawi moved less than 200 m. This finding throws into stark relief the importance of distinguishing between migration and mobility.

References

Abay, K., Asnake, W., Ayalew, H., Chamberlin, J. & Sumberg, J. 2020. Landscapes of opportunity: Patterns of young people's engagement with the rural economy in sub-Saharan Africa. *Journal of Development Studies* (doi. org/10.1080/00220388.2020.1808195).

Acosta, M., Van Wessel, M., Van Bommel, S., Ampaire, E. L., Twyman, J., Jassogne, L. & Feindt, P. H. 2019. What does it mean to make a 'joint' decision? Unpacking intra-household decision making in agriculture: Implications for policy and practice. *The Journal of Development Studies*, 1–20.

AfDB 2016. Africa's youth in the labour market. *In:* AfDB (ed.) *African Development Report 2015. Growth, Poverty and Inequality Nexus: Overcoming Barriers to Sustainable Development*, Abidjan, African Development Bank (AfDB).

African Union Commission 2006. *African Youth Charter*, Banjul, African Union Commission.

Ahmed, S. A., Cruz, M., Go, D. S., Maliszewska, M. & Osorio-Rodarte, I. 2016. How significant is sub-Saharan Africa's demographic dividend for its future growth and poverty reduction? *Review of Development Economics*, 20, 762–793.

Amanor, K. S. 2010. Family values, land sales and agricultural commodification in South-Eastern Ghana. *Africa: Journal of the International African Institute*, 80, 104–125.

Andersson Djurfeldt, A., Kalindi, A., Lindsjö, K. & Wamulume, M. 2019. Yearning to farm – Youth, agricultural intensification and land in Mkushi, Zambia. *Journal of Rural Studies*, 71, 85–93.

Ansell, N. 2018. *Education Systems, Aspiration and Learning in Remote Rural Settings*, London, Brunel University.

Anyidoho, N. A., Leavy, J. & Asenso-Okyere, K. 2012. Perceptions and aspirations: A case study of young people in Ghana's cocoa Sector. *IDS Bulletin*, 43, 20–32.

Arslan, A., Egger, E. M. & Winters, P. 2018. Migration, demography, and agri-food systems: Challenges and opportunities. *In:* Serraj, R. & Pingali, P. (eds.) *Agriculture and Food Systems to 2050 – Global Trends, Challenges and Opportunities*, Singapore, World Scientific Publishing Co. Pte. Ltd.

Baah-Boateng, W. 2016. The youth unemployment challenge in Africa: what are the drivers? *The Economic and Labour Relations Review*, 27, 413–431.

Beazley, H. 2015. Multiple identities, multiple realities: children who migrate independently for work in South-east Asia. *Children's Geographies*, 13, 296–309.

Bennell, P. 1996. Rates of return to education: Does the conventional pattern prevail in sub-Saharan Africa? *World Development*, 24, 183–199.

Berckmoes, L. & White, B. 2014. Youth, farming and precarity in rural Burundi. *European Journal of Development Research*, 26, 190–203.

Berckmoes, L. & White, B. 2016. Youth, farming and precarity in rural Burundi. *In:* Huijsmans, R. (ed.) *Generationing Development: A Relational Approach to Children, Youth and Development*, London, Palgrave Macmillan.

Bloom, D. E. & Williamson, J. G. 1998. Demographic transitions and economic miracles in emerging Asia. *The World Bank Economic Review*, 12, 419–455.

Bloom, D. E., Kuhn, M. & Prettner, K. 2017. Africa's prospects for enjoying a demographic dividend. *Journal of Demographic Economics*, 83, 63–76.

BMZ 2017. *One World – No Hunger: Future of the Rural World International G20 Conference, Berlin, 27–28 April 2017*, Bonn, German Ministry for Economic Cooperation and Development (BMZ).

Bourdieu, P. 1980. *The Logic of Practice,* Stanford, CA, Stanford University Press.

Bourdieu, P. 1984. *Distinction: A Social Critique of the Judgement of Taste*, London, Routledge.

Bourdieu, P. 1993. 'Youth' is just a word. *In:* Bourdieu, P. (ed.) *Sociology in Question*, London, Sage.

Bourdieu, P. & Passeron, J.-C. 1990. *Reproduction in Education, Society and Culture (L. Wacquant, Trans. 2 ed.),* London, Sage.

Brown, P., Lauder, H. & Ashton, D. 2011. *The Global Auction. The Broken Promises of Education, Jobs, and Incomes*, Oxford, UK, Oxford University Press.

Brück, T., Ferguson, N. T. N., Izzi, V. & Stojetz, W. 2016. *Jobs Aid Peace: A Review of the Theory and Practice of the Impact of Employment Programmes on Peace in Fragile and Conflict-affected Countries*, Berlin, International Security and Development Center (ISDC).

Carletto, C. & Gourlay, S. 2019. A thing of the past? Household surveys in a rapidly evolving (agricultural) data landscape: Insights from the LSMS-ISA. *Agricultural Economics*, 50, 51–62.

Carr, E. R. 2008. Men's crops and women's crops: The importance of gender to the understanding of agricultural and development outcomes in Ghana's Central Region. *World Development*, 36, 900–915.

Carter, M. R. 2016. What farmers want: the "gustibus multiplier" and other behavioral insights on agricultural development. *Agricultural Economics*, 47, 85–96.

Chamberlin, J. & Jayne, T. S. 2013. Unpacking the meaning of 'Market Access': Evidence from rural Kenya. *World Development*, 41, 245–264.

Chamberlin, J., Pender, J. & Yu, B. 2006. *Development Domains for Ethiopia: Capturing the Geographical Context of Smallholder Development Options*, Washington, DC, International Food Policy Institute (IFPRI).

Chant, S. & Jones, G. A. 2005. Youth, gender and livelihoods in West Africa: Perspectives from Ghana and the Gambia. *Children's Geographies*, 3, 185–199.

Christiaensen, L. 2017. Agriculture in Africa – Telling myths from facts: a synthesis. *Food Policy*, 67, 1–11.

Christiaensen, L. & Demery, L. (eds.) 2018. *Agriculture in Africa: Telling Myths from Facts*, Washington, DC, The World Bank.

Christiansen, C., Utas, M. & Vigh, H. (eds.) 2006. *Navigating Youth, Generating Adulthood: Social Becoming in an African Context*, Stockholm, Elanders Gotab AB.

Cole, J. 2011. A cultural dialectics of generational change. The view from contemporary Africa. *Review of Research in Education*, 35, 60–88.

De Wilde, R. 2000. *De Voorspellers: Een Kritiek Op De Toekomstindustrie*, Amsterdam, De Balie.

Dercon, S. 2002. Income risk, coping strategies, and safety nets. *The World Bank Research Observer*, 17, 141–166.

Dercon, S. & Krishnan, P. 2000. In sickness and in health: Risk sharing within households in rural Ethiopia. *Journal of Political Economy*, 108, 688–727.

Dhillon, N. & Yousef, T. (eds.) 2009. *Generation in Waiting: The Unfulfilled Promise of Young People in the Middle East*, Washington, DC, Brookings Institution Press.

Doss, C., Kovarik, C., Peterman, A., Quisumbing, A. & Van Den Bold, M. 2015. Gender inequalities in ownership and control of land in Africa: myth and reality. *Agricultural Economics*, 46, 403–434.

Drummond, P., Thakoor, V. & Yu, S. 2014. *Africa Rising: Harnessing the Demographic Dividend. IMF Working Paper WP/14/143*, Washington, DC, International Monetary Fund (IMF).

Dunne, M. 2007. Gender, sexuality and schooling: Everyday life in junior secondary schools in Botswana and Ghana. *International Journal of Educational Development*, 27, 499–511.

Dunne, M. & Adzahlie-Mensah, V. 2015. Subordinate subjects: The work of the hidden curriculum in post-colonial Ghana. *In:* Wyse, D., Hayward, L. & J., P. (eds.) *The SAGE Handbook of Curriculum, Pedagogy and Assessment*, London, Sage.

Eastwood, R. & Lipton, M. 2011. Demographic transition in sub-Saharan Africa: How big will the economic dividend be? *Population Studies-a Journal of Demography*, 65, 9–35.

Elias, M., Mudege, N., Lopez, D. E., Najjar, D., Kandiwa, V., Luis, J., Yila, J., Tegbaru, A., Ibrahim, G., Badstue, L., Njuguna-Mungai, E. & Bentaibi, A. 2018. Gendered aspirations and occupational trajectories among rural youth: A cross-regional perspective. *Journal of Gender, Agriculture and Food Security*, 3, 82–107.

Erikson, M. G. 2007. The meaning of the future: Toward a more specific definition of possible selves. *Review of General Psychology*, 11, 348–358.

FAO 2015. *The State of Food and Agriculture: Migration and Rural Development*, Rome, Food and Agriculture Organization of the United Nations (FAO).

Filmer, D. & Fox, L. 2014. *Youth Employment in Sub-Saharan Africa*, Washington, DC, Agence Française de Développement and The World Bank.

Flachs, A. & Richards, P. 2018. Playing development roles: the political ecology of performance in agricultural development. *Journal of Political Ecology*, 25, 638–646.

Flynn, J. & Sumberg, J. 2017. Youth savings groups in Africa: they're a family affair. *Enterprise Development and Microfinance*, 28, 147–161.

Glover, D. & Sumberg, J. 2020. Youth and food systems transformation. *Frontiers in Sustainable Food Systems*, 4 (doi.org/10.3389/fsufs.2020.00101).

Hardgrove, A., Rootham, E. & Mcdowell, L. 2015. Possible selves in a precarious labour market: youth, imagined futures, and transitions to work in the UK. *Geoforum*, 60, 163–171.

Hashim, I. 2005. *Exploring the Linkages Between Children's Independent Migration and Education: Evidence from Ghana*, Brighton, UK, Sussex Centre for Migration Research, University of Sussex.

Hashim, I. & Thorsen, D. 2011. *Child Migration in Africa*, New York, Zed Books.

Honwana, A. 2012. *The Time of Youth: Work, Social Change, and Politics in Africa*, Boulder, CO, Kumarian Press.

Honwana, A. 2019. Youth struggles: From the Arab Spring to Black Lives Matter & beyond. *African Studies Review*, 62, 8–21.

Hopkins, P. & Pain, R. 2007. Geographies of age: thinking relationally. *Area*, 39, 287–294.

Huijsmans, R. (ed.) 2016. *Generationing Development: A Relational Approach to Children, Youth and Development*, London, Palgrave Macmillan.

Humphreys, S., Undie, C.-C. & Dunne, M. 2008. Gender, sexuality and development: Key issues in education and society in sub-Saharan Africa. *In:* Dunne, M. (ed.) *Gender, Sexuality and Development. Education and Society in sub-Saharan Africa*. Rotterdam, Sense Publishers.

IFAD 2019. *Creating Opportunities for Rural Youth: 2019 Rural Development Report*, Rome, International Fund for Agricultural Development (IFAD).

ILO 2015. *Global Employment Trends for Youth 2015: Scaling Up Investments in Decent Jobs for Youth*, Geneva, International Labour Organization (ILO).

Jacquemin, M. 2012. *'Petites bonnes' d'Abidjan. Sociologie des Filles en Service Domestique*, Paris, L'Harmattan.

Jansen, K. & Gupta, A. 2009. Anticipating the future: 'Biotechnology for the poor' as unrealized promise? *Futures*, 41, 436–445.

Jedwab, R., Christiaensen, L. & Gindelsky, M. 2017. Demography, urbanization and development: Rural push, urban pull and...urban push? *Journal of Urban Economics*, 98, 6–16.
Jeffrey, C. 2010. Timepass: youth, class, and time among unemployed young men in India. *American Ethnologist*, 37, 465–481.
Jones, M. D. & Mcbeth, M. K. 2010. A narrative policy framework: Clear enough to be wrong? *Policy Studies Journal*, 38, 329–353.
Kleist, N. & Thorsen, D. (eds.) 2017. *Hope and Uncertainty in Contemporary African Migration*, New York and London, Taylor & Francis.
Kosec, K., Ghebru, H., Holtemeyer, B., Mueller, V. & Schmidt, E. 2016. *The Effect of Land Inheritance on Youth Employment and Migration Decisions: Evidence From Rural Ethiopia*, Washington, DC, International Food Policy Institute (IFPRI).
Kosec, K., Ghebru, H., Holtemeyer, B., Mueller, V. & Schmidt, E. 2018. The effect of land access on youth employment and migration decisions: evidence from rural Ethiopia. *American Journal of Agricultural Economics*, 100, 931–954.
Kremer, M., Rao, G. & Schilbach, F. 2019. Behavioral development economics. *In:* Bernheim, D., Dellavigna, S. & Laibson, D. (eds.) *Handbook of Behavioral Economics: Applications and Foundations 2*. Amsterdam, North Holland.
Langevang, T. 2008. 'We are managing!' uncertain paths to respectable adulthoods in Accra, Ghana. *Geoforum*, 29, 2039–2047.
Leavy, J. & Hossain, N. 2014. *Who Wants to Farm? Youth Aspirations, Opportunities and Rising Food Prices. IDS Working Paper 439*, Brighton, UK, Institute of Development Studies (IDS).
Lesclingand, M. & Hertrich, V. 2017. Quand les filles donnent le ton. Migrations adolescentes au Mali. *Population*, 72, 63–93.
Linares, O. F. 2003. Going to the city... and coming back? Turnaround migration among the Jola of Senegal. *Africa: Journal of the International African Institute*, 73, 113–132.
Losch, B. 2016. *Structural Transformation to Boost Youth Labour Demand in Sub-Saharan Africa: The Role of Agriculture, Rural Areas and Territorial Development. Employment Working Paper No. 204*, Geneva, International Labour Organization (ILO).
Mabiso, A. & Benfica, R. 2019. *The Narrative on Rural Youth and Economic Opportunities in Africa: Facts, Myths and Gaps. IFAD Research Series 61*, Rome, International Fund for Agricultural Development (IFAD).
Mannheim, K. 1952. The problem of generations. *In:* Kecskemeti, P. (ed.) *Essays on the Sociology of Knowledge: Collected Works, Volume 5*. New York, Routledge.
Marginson, S. 2019. Limitations of human capital theory. *Studies in Higher Education*, 44, 287–301.
Markus, H. & Nurius, P. 1986. Possible selves. *American Psychologist*, 41, 954–969.
Mberu, B. U. 2005. Who moves and who stays? Rural out-migration in Nigeria. *Journal of Population Research*, 22, 141–161.
McGrath, S., Ramsarup, P., Zeelen, J., Wedekind, V., Allais, S., Lotz-Sisitka, H., Monk, D., Openjuru, G. & Russon, J.-A. 2019. Vocational education and training for African development: a literature review. *Journal of Vocational Education & Training*, 1–23.
Mueller, V. & Lee, H. L. 2019. Can migration be a conduit for transformative youth employment? *In:* Mueller, V. & Thurlow, J. (eds.) *Youth and Jobs in Rural Africa: Beyond Stylized Facts*, Oxford, UK, Oxford University Press.
Mueller, V. & Thurlow, J. (eds.) 2019. *Youth and Jobs in Rural Africa: Beyond Stylized Facts*, Oxford, UK, Oxford University Press.
Mueller, V., Thurlow, J., Rosenbach, G. & Masias, I. 2019. Africa's rural youth in the global context. *In:* Mueller, V. & Thurlow, J. (eds.) *Youth and Jobs in Rural Africa: Beyond Stylized Facts*, Oxford, UK, Oxford University Press.
Navarro, Z. 2006. In search of cultural interpretation of power. *IDS Bulletin*, 37, 11–22.
OECD 2017. *Youth Aspirations and the Reality of Jobs in Developing Countries: Mind the Gap*, Paris, Development Centre Studies, OECD Publishing.
Pender, J., Jagger, P., Nkonya, E. & Sserunkuuma, D. 2004. Development pathways and land management in Uganda. *World Development*, 32, 767–792.
Pender, J., Place, F. & Ehui, S. 2006. *Strategies for Sustainable Land Management in the East African Highlands*, Washington, DC, International Food Policy Institute (IFPI).
Petesch, P. & Rodríguez Caillava, I. 2012. *Voices of Young Villagers in Sub-Saharan Africa. Background paper*, Washington, DC, The World Bank.

Pettit, J. 2016. Why citizens don't engage. Power, poverty and civic habitus. *IDS Bulletin*, 47, 89–102.

Phillips, A. 2010. What's wrong with essentialism? *Distinktion: Journal of Social Theory*, 11, 47–60.

Porter, G., Hampshire, K., Abane, A., Robson, E., Munthali, A., Mashiri, M. & Tanle, A. 2010. Moving young lives: Mobility, immobility and inter-generational tensions in urban Africa. *Geoforum*, 41, 796–804.

Potts, D. 2013. *Rural-urban and Urban-rural Migration Flows as Indicators of Economic Opportunity in Sub-Saharan Africa: What Do the Data Tell Us? Working Paper 9*, Brighton, UK, Migrating Out of Poverty Research Programme Consortium.

Republic of Kenya 2007. *National Youth Policy for Youth Development*, Nairobi, Republic of Kenya.

Republic of Malawi 2013. *National Youth Policy*, Lilongwe, Republic of Malawi, Ministry of Youth and Sports.

Richards, P. 1986. *Coping With Hunger: Hazard and Experiment in an African Rice Farming System*, London, Allen & Unwin.

Richards, P. 1989. Agriculture as a performance. *In:* Chambers, R., Pacey, A. & Thrupp, L. A. (ed.) *Farmer First: Farmer Innovation and Agricultural Research*, London, Intermediate Technology Publications.

Roberts, K. 1968. The entry into employment: an approach towards a general theory. *Sociological Review*, 16, 165–184.

Roberts, K. 1977. The social conditions, consequences and limitations of careers guidance. *British Journal of Guidance & Counselling*, 5, 1–9.

Roberts, K. 2009. Opportunity structures then and now. *Journal of Education and Work*, 22, 355–368.

Roe, E. M. 1991. Development narratives, or making the best of blueprint development. *World Development*, 19, 287–300.

Roe, E. M. 1995. Except Africa postscript to a special section on development narratives. *World Development*, 23, 1065–1069.

Ruiz Salvago, M., Phiboon, K., Faysse, N. & Nguyen, T. P. L. 2019. Young people's willingness to farm under present and improved conditions in Thailand. *Outlook on Agriculture*, 48, 282–291.

Scoones, I., Mavedzenge, B. & Murimbarimba, F. 2019. Young people and land in Zimbabwe: livelihood challenges after land reform. *Review of African Political Economy*, 46, 117–134.

Seymour, G. & Peterman, A. 2018. Context and measurement: An analysis of the relationship between intrahousehold decision making and autonomy. *World Development*, 111, 97–112.

Shanahan, E. A., Jones, M. D. & Mcbeth, M. K. 2011. Policy narratives and policy processes. *Policy Studies Journal*, 39, 535–561.

Singerman, D. 2007. *The Economic Imperatives of Marriage: Emerging Practices and Identities Among Youth in the Middle East. Working Paper 6,* Washington, DC and Dubai, Wolfensohn Centre for Development and Dubai School of Government.

Sommers, M. 2012. *Stuck: Rwandan Youth and the Struggle for Adulthood*, Atlanta, University of Georgia Press, in association with US Institute of Peace Press.

Streletskaya, N. A., Bell, S. D., Kecinski, M., Li, T., Banerjee, S., Palm-Forster, L. H. & Pannell, D. 2020. Agricultural adoption and behavioral economics: Bridging the gap. *Applied Economic Perspectives and Policy*, 42, 54–66.

Sumberg, J. & Hunt, S. 2019. Are African rural youth innovative? Claims, evidence and implications. *Journal of Rural Studies*, 69, 130–136.

Sumberg, J., Glover, D., Chamberlin, J., Flynn, J. & Johnson, V. 2019. *Landscapes of Rural Youth Opportunity. IFAD Research Series 47*, Rome, IFAD.

Sumberg, J., Flynn, J., Fox, L., Mader, P. & Oosterom, M. 2021. Africa's 'youth employment' crisis is actually a 'missing jobs' crisis. *Development Policy Review* (doi.org/10.1111/dpr.12528).

Tadele, G. & Gella, A. A. 2012. 'A last resort and often not an option at all': Farming and young people in Ethiopia. *IDS Bulletin*, 43, 33–34.

Temudo, M. & Abrantes, M. 2015. The pen and the plough: Balanta young men in Guinea-Bissau. *Development and Change*, 46, 464–485.

The World Bank 2014. *Aspiration Traps: When Poverty Stifles Hope*, Washington, DC, The World Bank.

Thorsen, D. 2013. Weaving in and out of employment and self-employment: young rural migrants in the informal economy of Ouagadougou. *International Development Planning Review*, 35, 203–218.

Timmer, C. P. 1988. Chapter 8. The agricultural transformation. *In:* Chenery, H. & Srinivasan, T. N. (eds.) *Handbook of Development Economics, Vol. 1*. Amsterdam and New York, North Holland / Elsevier.

UNDESA n.d. *Definition of Youth*, New York, United Nations Department of Economic and Social Affairs (UNDESA).
UNEP 2019. *GEO-6 for Youth Africa: A Wealth of Green Opportunities*, Nairobi, United Nations Environment Programme (UNEP).
UNESCO 2015. *Global Monitoring Report 2000–2015 Achievements and Challenges*, Paris, UNESCO.
UNESCO 2018. *Global Education Monitoring Report. Accountability on Re-education: Meeting Our Commitment,* Paris, UNESCO.
UNFPA 2014. *The State of World Population 2014. The Power of 1.8 Billion: Adolescents, Youth and the Transformation of the Future*, New York, United Nations Population Fund (UNFPA).
Van Blerk, L. 2016. Livelihoods as relational im/mobilities: Exploring the everyday practices of young female sex workers in Ethiopia. *Annals of the American Association of Geographers*, 106, 413–421.
Van Den Broeck, G. & Kilic, T. 2019. Dynamics of off-farm employment in sub-Saharan Africa: A gender perspective. *World Development*, 119, 81–99.
Van Geel, J. & Mazzucato, V. 2018. Conceptualising youth mobility trajectories: thinking beyond conventional categories. *Journal of Ethnic and Migration Studies*, 44, 2144–2162.
Vigh, H. 2006. Social death and violent life chances. *In:* Christiansen, C., Utas, M. & Vigh, H. (eds.) *Navigating Youth, Generating Adulthood: Social Becoming in an African Context*, Uppsala, Sweden, Nordiska Afrikainstitutet.
Vigh, H. 2009. Motion squared: a second look at the concept of social navigation. *Anthropological Theory*, 9, 419–438.
White, B. 2015. Generational dynamics in agriculture: reflections on rural youth and farming futures. *Cahiers Agricultures*, 24, 330–334.
Wiggins, S. & Proctor, S. 2001. How special are rural areas? The economic implications of location for rural development. *Development Policy Review*, 19, 427–436.
Yazbeck, A. S., Raja, S. & Canning, D. 2015. *Africa's Demographic Transition*, Washington, DC, The World Bank and Agence Française de Développement.
Yeboah, T., Sumberg, J., Flynn, J. & Anyidoho, N. A. 2017. Perspectives on desirable work: Findings from a Q study with students and parents in rural Ghana. *European Journal of Development Research*, 29, 423–440.
Yeboah, T., Chigumira, E., John, I., Anyidoho, N. A., Manyong, V., Flynn, J. & Sumberg, J. 2020. Hard work and hazard: Young people and agricultural commercialisation in Africa. *Journal of Rural Studies*, 76, 142–151.
Zipin, L., Sellar, S., Brennan, M. & Gale, T. 2015. Educating for futures in marginalized regions: a sociological framework for rethinking and researching aspirations. *Educational Philosophy and Theory: Incorporating ACCESS*, 47, 227–246.

2 Empirical Windows on African Rural Youth

Marjoke Oosterom[1], Jordan Chamberlin[2] and James Sumberg[1]
[1]*Institute of Development Studies, Brighton, UK;* [2]*International Maize and Wheat Improvement Centre (CIMMYT), Nairobi, Kenya*

Introduction

Over the last two decades young people in rural areas of sub-Saharan Africa (SSA) have become a clear focus for political rhetoric, policy and investment at national, regional and international levels. Policy processes and public discourse addressing rural youth, and the arguments that inform them, draw on a wide array of data and information sources. These range from official statistics (e.g. unemployment and school enrolment rates), through nationally representative surveys (e.g. the Living Standards Measurement Study – Integrated Surveys on Agriculture (LSMS–ISA) programme of The World Bank (Carletto and Gourlay, 2019)), special purpose surveys and mixed method or qualitative studies at selected sites. The field is also rife with 'common knowledge' claims and narratives about 'African youth': for instance, that they are abandoning agriculture and fleeing rural areas or that they are particularly innovative (for further discussion of these narratives see Mabiso and Benfica, 2019; Sumberg and Hunt, 2019). While many of these stylized claims are not rooted in empirical evidence, they nevertheless continue to play important roles in public and policy discourse.

In many respects, attempting to say anything meaningful and 'evidence-based' about 'young people in rural Africa' or even 'young people in country X' is a herculean task, if not altogether hopeless. One is immediately confronted with the fact of high levels of diversity among both young people and rural areas. The obvious riposte to any finding or statement about young people in rural Africa is simply, 'Which young people, and which rural Africa?'

The alternative available sources of empirical information provide different kinds of insights and differ in how effectively they deal with the challenge of 'which young people, and which rural Africa?'. For example, large-scale survey data have the appeal of enabling statistical inference around measurable outcomes (e.g. the share of 17-year-olds in school in a country), but may be wholly insufficient at representing subjective views of educational quality, constraints to staying in school, plans for the future, or other nuanced descriptions of situations more discoverable through qualitative methods. In summarizing the conclusions drawn from different empirical sources, the strengths and weaknesses of alternative methodologies and evidence bases are not often explored, or even acknowledged, at least in relation to how young people engage with the rural economy. Further, there has been little discussion of how these respective strengths and weaknesses might affect understandings of young people's economic activities, their efforts to build their livelihoods, and hence what policy and intervention strategies would be most appropriate.

DOI: 10.1079/9781789245011.0002

In this chapter we review a range of academic studies that are centrally focused on or include findings about young people and the rural economy in Africa. We have selected a number of studies that have used quantitative, qualitative and mixed method approaches and employ these to reflect on the question: 'How do we know about young people in rural SSA?' Specifically, the strengths and weaknesses of the main methodological approaches used to generate empirical evidence about how young people engage with the rural economy are analysed. As we will show, the differences across the research literature are significant. For example, some studies address youth-related questions by extracting data on the activities of rural youth from general household surveys, while others are designed to specifically involve youth as research participants. Studies about rural youth have different theoretical, conceptual and disciplinary underpinnings, which are then reflected in coverage (e.g. national, multi-sited or single-sited), sample size and sampling strategy. Some large-scale studies are cross sectional while others are longitudinal; other analyses are highly contextualized case studies. There are also differences in the primary data collection unit (a household or an individual young person), the methods (quantitative, qualitative or mixed) and whether or not the study has an ambition to engage with policy.

What we will demonstrate in this chapter, is that even multi-sited case studies do not always investigate how particularities of geographic locations influence young people's economic activities. Nor does social diversity among rural youth often figure prominently in most analyses, despite the avowed desire to avoid treating African 'youth' as a homogeneous group. Perhaps the most common way of addressing diversity is by using a gender lens. However, while undoubtedly important, gender remains only one aspect of social difference, and gender differences may or may not map on to differences in age (even among the 'youth'), ethnicity, religion, disability, wealth, level of education or origin (indigene or migrant). Even if different social categories of young people have been purposively included in the sample, the implications of social difference – for example, for accessing educational and work opportunities – are seldom explored in depth. It is here that the ideal of 'policy-relevant research' becomes unstuck: while policy makers want simple facts and narratives covering large numbers of rural youth, these same facts and narratives effectively extract young people from their social, economic and geographical contexts. Yet it is these social and geographical contexts that enable and constrain young people's economic activities, allowing some to move ahead while disadvantaging others. Different research approaches can, in principle, deal with context – for example, by 'controlling for it' or building it into the design of empirical work. To date, however, the role of context remains a major lacuna in the evidence base on the economic activities of rural youth.

The remainder of the chapter is organized in three sections. The next section focuses on the different empirical windows that have been used to study young people's economic lives and livelihoods. A selection of studies is analysed in terms of the questions addressed, methods used and the scale of the spatial analysis. Following this, the methods employed in the research presented in this book are described and situated in relation to the larger body of literature touching on youth and the rural economy. The final section makes some specific recommendations on how these windows and methods might be improved to gain greater, more policy-relevant insight on young people's lives and livelihoods, in all their diversity.

Empirical Windows: How Do We Know About Rural Youth in Africa?

To frame this exercise, we generated a list of empirical studies that directly or indirectly address youth engagement with the rural economy in SSA. This collection of 39 studies (Appendix Table 2A) reflects our own knowledge of the literature and suggestions received from several other researchers in the field. It is not meant to be exhaustive (Francophone and Lusophone literature is absent), but rather to reflect both the essence and the diversity of the empirical literature. Included in Appendix Table 2A are the 'cited by' figures given by Google Scholar, which, with all the usual caveats (i.e. difference in publication date and type of publication), provide a crude indicator of the relative influence of the different studies to date.[1]

Each study was then broadly classified in relation to the methods used (primarily quantitative, primarily qualitative or mixed) and the scale at which the analysis is focused (Fig. 2.1). The boundaries between quantitative and qualitative

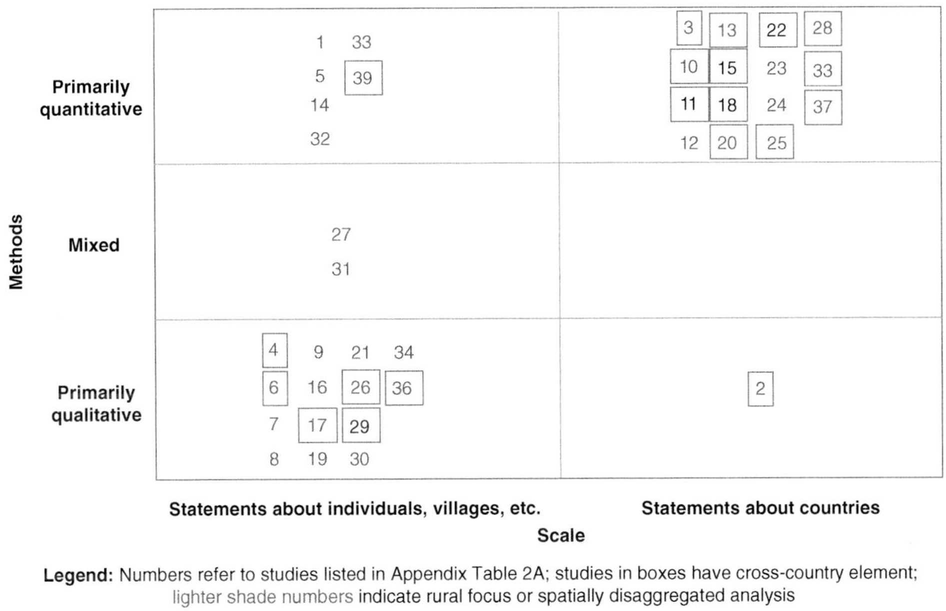

Fig. 2.1. Distribution of selected empirical studies by method and scale. Courtesy of the authors.

are not absolute, and some quantitative studies have a qualitative element and vice versa. The two largest clusters of studies are first, those that use primarily quantitative methods and representative samples to present analyses at the level of a whole country; and second, those that use qualitative methods to present analyses at the level of individuals, villages or districts (although most often these are not based on representative samples). Most studies in the former group have a cross-country comparative element, while fewer studies in the latter group are concerned with country comparisons. It is also important to note that relatively highly cited studies appear in four of the six boxes, suggesting there is no clear methodological monopoly on the knowledge about young people's engagement with the rural economy.

In the sections that follow we review many of these studies in more detail, with a focus on the topics they address, the methods used, and briefly, whether they support or challenge common narratives about rural youth.

Studies with a primarily quantitative orientation

The quantitative studies overwhelmingly address three major topics: employment (Blattman *et al.*, 2014; Filmer and Fox, 2014; Elder *et al.*, 2015; Maiga *et al.*, 2015; Baah-Boateng, 2016; Fox and Thomas, 2016; Losch, 2016; Yeboah and Jayne, 2018; Mueller and Thurlow, 2019, Chapters 2, 3, 5, 6, 7, 8 and 9); land access (Kristensen and Birch-Thomsen, 2013; Bezu and Holden, 2014; Kosec *et al.*, 2018; Andersson Djurfeldt *et al.*, 2019); and migration and labour mobility (Bezu and Holden, 2014; Kosec *et al.*, 2018; Mueller and Thurlow, 2019, Chapters 2 and 9).

These topics are linked in ways that largely align with two major themes. The first is that farming opportunities are increasingly limited for young people in SSA – chiefly, but not exclusively, because their access to land is limited. The second theme is that non-farm income-generating opportunities available to young people are also limited in rural (and urban) areas. The upshot is that the lack of economic opportunity will drive a number of processes with either negative or ambiguous valence, including rural out-migration, political unrest, and fraying of traditional social norms and institutions.[ii]

Many of the studies in our selection focus on descriptive documentation of youth employment patterns. A central question is the relative importance of agriculture as an employer of rural youth. Given the limited absorption capacity of urban areas, and the small non-farm sector in many rural areas, agriculture is widely seen as having a central role in the provision of productive employment for Africa's rural youth (Filmer and Fox, 2014; AGRA, 2015; Maiga *et al.*, 2015; Losch, 2016; IFAD, 2019). Maiga *et al.* (2015) estimate agricultural labour participation rates for youth (defined as individuals aged 16–35) in Ethiopia, Malawi, Niger, Nigeria, Tanzania and Uganda, finding about half of all youth, on average, worked in agriculture (although with high cross-country variation – rates ranged from 27.1% in Nigeria to 63.4% in Niger – combined with high internal variation). Yeboah and Jayne (2018), using nationally representative data from nine countries, find that the number of young people engaged in farming is increasing in absolute terms but the share of farming in total employment is declining over time in most countries. Employment in off-farm segments of agrifood systems is expanding rapidly in percentage terms, with non-farm activities constituting the major source of new employment. This movement of labour from farming is associated with agriculture-driven changes: the performance of farming is found to significantly influence the rate of job growth in the rest of the economy, with a 5-year lagged moving average rate of agricultural productivity growth found to be positively and significantly associated with growth in the off-farm share of labour.

Other country-level studies have produced results which suggest that these patterns vary considerably across settings. Schmidt and Bekele Woldeyes (2019) find that young Ethiopians have a greater probability of working in non-farm enterprises than older individuals, but that outcomes are also affected by other ('push') factors: those in areas with less favourable agricultural potential, who possess fewer assets, and have less access to agricultural credit are more likely to seek off-farm work. Diao *et al.* (2019) find continuous movement of young people in Ghana from agriculture. Adesugba and Mavrotas (2016) find very large internal migration rates in Nigeria, much of which is associated with non-farm employment, particularly for those with higher educational attainments. On the other hand, Benson *et al.* (2019) find that the process of rural labour diversification into non-farm sectors in Malawi is not youth driven, but rather characterized by older men (in their 30s and 40s) responding to climate stresses and other push factors. They furthermore find little evidence of structural transformation processes alluded to in many of the large-scale assessments of SSA. These latter two studies find a strong association between non-farm work and remoteness, as does some of the background work for IFAD (2019).

In one of the few quantitative studies to examine farm decisions made by youth, Diao *et al.* (2019) find that compared with older farmers, younger farmers in Ghana are not more likely to intensify or use modern inputs; constraints to adoption seem to be non-age specific (e.g. market access is important for technology adoption) (see Chapter 6, this volume). They also find that labour productivity is growing faster than land productivity – highlighting the increasing importance of labour-saving technologies for agricultural intensification. How this will play out for successive generations of new entrants to farming is unclear, given the land and migration patterns mentioned above.

Andersson Djurfeldt *et al.*'s (2019) mixed methods study in Zambia also found no age-related intensification or productivity relationships, which they attribute in part to the limited resources of younger household heads – a finding which recurs repeatedly in both quantitative and qualitative assessments. Land was found to be a major constraint to young people's farming prospects, as elsewhere – e.g. Andriamanalina *et al.* (2014) in Madagascar, Ahaibwe *et al.* (2013) in Uganda, Benson *et al.* (2019) in Malawi.

A recurring finding is that, because of limited rural opportunities (again linked with underdeveloped land and labour markets), the movement of young people from rural areas is increasing and likely to increase further over time. Bezu and Holden (2014) and Kosec *et al.* (2018) both use microeconomic data from subnational samples in highland Ethiopia and find that youth out-migration is in part a response to limited agricultural land availability. Mueller and Lee's (2019) Ethiopian study finds that rural-to-urban migration facilitates movement out of agriculture, with associated gains in economic returns, but it is a pathway for relatively few young people. Youth are more likely to pursue rural-to-rural migration.

The political and developmental valence of this movement is less clear: on the one hand, given limited absorption capacities of the non-farm sectors in both urban and rural areas, such flows will not address the youth employment challenge (Filmer and Fox, 2014; Losch, 2016). On the other hand, economic welfare returns to investment in movement are generally perceived to be high, particularly for those able to move to urban areas (Mueller and Lee, 2019).

The analytical emphasis of many quantitative studies is on identifying the determinants of observable outcomes, where the determinants of interest are often structural features of the economic setting which may be amenable to policy leverage (e.g. functioning of land markets). Emphasis is on formal conceptualization of relationships, identification of average effects at a population level, and credibility of claims for internal validity (i.e. claims of causal identification). Most of the studies in the set of papers that use quantitative methods feature individual- or household-level observational data and econometric analyses (e.g. Bezu and Holden, 2014; Maiga *et al.*, 2015; Baah-Boateng, 2016; Kosec *et al.*, 2018; Yeboah and Jayne, 2018; Andersson Djurfeldt *et al.*, 2019; Mueller and Thurlow, 2019, most chapters). Others feature macroeconomic analyses to make statements about structural relationships at the level of the national economy (Filmer and Fox, 2014; Baah-Boateng, 2016; Fox and Thomas, 2016; Losch, 2016).[iii] Most of these studies seek to identify average treatment effects or marginal effects of changes in variables that can be altered (in theory) through policy.

A key feature of many of these studies is their large sample sizes and sampling strategies that make then nationally representative.[iv] Internal diversity in such large, random samples is almost guaranteed, by construction. However, policy-oriented analysis typically seeks to uncover average effects after controlling for such diversity, rather than to explicitly examine its implications for policy, e.g. through interaction effects of key characteristics, such as age and gender, or particular contexts, such as remote rural settings. Relatedly, comparative analyses of different geographical contexts are only presented in a few of the studies, as in many cases the extent of potential disaggregation is limited by sample design. Surprisingly, educational attainment of young people – a key component of policy debates – is not the central focus of any of the studies in our selected set, although limited education is cited as a broad constraint on the non-farm opportunities available to rural young people (e.g. Yeboah and Jayne, 2018) and for those who migrate to urban areas (Mueller and Thurlow, 2019, Chapter 2). While sex disaggregated analysis features in many of these studies (e.g. Bezu and Holden, 2014; Kosec *et al.*, 2018; Yeboah and Jayne, 2018; Andersson Djurfeldt *et al.*, 2019), further interactions (e.g. between sex and educational attainment) are generally not addressed and gender-specific policy recommendations are not made. Policy prescriptions tend to be very normative and fail to articulate ways in which youth as a heterogenous group may require differentiated support; this may encourage essentialist thinking and overly standardized policy actions (Sumberg *et al.*, 2012).

The scale of analysis is a function of the sampling frames, with studies that use nationally representative data generally making inferential claims about all rural youth in a country (e.g. the studies in Mueller and Thurlow, 2019). Claims for external validity elsewhere in the region are not usually made explicitly, except in the case of cross-country analyses (Maiga *et al.*, 2015; Yeboah and Jayne, 2018). However, the fact of national representativeness confers a strong inherent advantage of such studies over stocktaking claims arising from qualitative case studies which are typically much reduced in scope. Some subnational surveys typically offer large samples (many hundreds of observations) which are either representative of key subpopulations of interest or indicative of broadly shared conditions (e.g. *woredas* within the Blue Nile Basin in Amhara and Oromia, Ethiopia, see Kosec *et al.*, 2018).

Unlike qualitative studies, most quantitative analyses make statements about young people using data sets collected via multi-purpose household-level instruments (although typically with detailed information on individual members, reported by a single respondent, usually the head). Youth-specific instruments are rare (exceptions include Bezu and Holden, 2014; Kosec *et al.*, 2018; Andersson Djurfeldt *et al.*, 2019). Thus, data are generally collected on – but not from – young people via the information gathered for members of sampled households (e.g. Maiga *et al.*, 2015; Yeboah and Jayne, 2018; Mueller and Thurlow, 2019, Chapters 2, 3, 5, 6, 7, 8 and 9). The fact that data underpinning these analyses are generally not reported by the individuals concerned is potentially problematic, as the head (the typical respondent in household surveys) may not be aware of or adequately represent the activities and experiences of young household members (see Chapter 6, this volume). There is a large empirical literature documenting divergence of responses to the same questions from self-reporting versus proxy respondents, and gendered differences in estimates of asset values, control and decision making, and time spent under different activities (Bardasi *et al.*, 2011; Kilic and Moylan, 2016; Doss *et al.*, 2017, 2018; Janzen, 2018; Ambler *et al.*, 2019).

As noted above, the emphasis of many of these studies is to gain insights about the outcomes that might be expected if conditions are changed through policy or through specific interventions. This orientation can generate valuable inputs for policy guidance. However, such studies generally do not directly address processes, path dependencies or dynamics, other than through the interpretive discussion of descriptive statistics and estimated relationships. For the most part, transitions to adulthood are not directly addressed, except in some interpretive discussions of results (e.g. Andersson Djurfeldt *et al.*, 2019). Although many studies use panel data (in which the same individuals and/or households are observed over time), these data are used to control for unobserved heterogeneity in order to make better appeals for internal validity. To date, panel data have not been used to describe the trajectories of young women's and men's engagement with the rural economy. It may be useful to pursue analysis along these lines in the future.

Studies with a primarily qualitative orientation

The studies discussed in this section can be divided into two groups. The first, including Petesch and Rodríguez Caillava (2012), Leavy and Hossain (2014) and Yeboah *et al.* (2020) are relatively large scale, including many respondents and multiple locations and/or countries. Petesch and Rodríguez Caillava (2012) report on work designed to provide background to the 2012 World Development Report on Gender Equality and Development (The World Bank, 2012). The study focuses on people's perceptions of economic opportunities, how they approached these opportunities, and the social institutions that shaped their perceptions. It integrates a gender perspective in the choice of topics and questions, addressing issues such as caring responsibilities and marriage traditions. Data were collected through nearly 150 focus groups and key informant interviews, in a total of 30 rural communities across six countries in SSA (Burkina Faso, Liberia, South Africa, Sudan, Tanzania and Togo). Focus groups were conducted separately with young women and men from ages 18 to 24, as well as with adult women and men (ages 25 to 60). In Burkina Faso, Sudan and Togo, additional focus groups were held with adolescent boys and girls (ages 10 to 16).

The study by Leavy and Hossain (2014) is extracted from a larger research programme on food price volatility in the wake of the 2008 financial crisis. Using the concept of 'opportunity space' (Sumberg *et al.*, 2012), it aims to provide insights into the opportunities for youth to engage with farming. The study explores the assumption that higher food prices make farming relatively more attractive. Using a sociological approach, it examines the factors influencing choices with respect to farming and how aspirations are formed. In total, 13 rural sites and ten peri-urban and urban sites in ten countries were covered (Vietnam, Bangladesh, Indonesia, Pakistan, Ethiopia, Zambia, Burkina Faso, Kenya, Guatemala and Bolivia). Focus group discussions (FDGs) were held in all sites with individuals in the low-income category or in precarious livelihoods, while individual interviews were conducted with youth, parents and older key informants. In total 1500 participants took part in the study (the number of FGDs and interviews is not specified). Yeboah *et al.* (2020) is a medium-scale, multi-country study focused on young people's engagement in farming and other economic activities in 'hot spots' of agricultural commercialization. Data were collected via 117 interviews conducted in single sites in Tanzania, Zimbabwe and Ghana with young women and men aged between 16 and 35. The strategy for capturing diversity was to have a gender-balanced sample, and including individuals of different ages, origin (local or migrant) and years of formal education.

The second group of qualitative studies includes smaller-scale studies involving one or a few locations (Ansell, 2004; Locke and te Lintelo, 2012; Kristensen and Birch-Thomsen, 2013; Berckmoes and White, 2014; Reynolds-Whyte and Acio, 2017). These seek mainly to contribute to conceptual development, and although there may be some engagement with policy narratives and discourses, the primary objective is not to provide policy or programme recommendations. The studies explore processes and dynamics and seek to explain outcomes such as young people's educational attainment, economic activity and transitions to adulthood.

Ansell (2004) investigates the gendered impact of schooling on young people's transitions, and the multiple factors that influence their secondary school experiences and hopes for the future. In terms of scale this was the smallest study in our set: case study research was conducted at two rural secondary schools, one in Lesotho and one in Zimbabwe, which included participant observation, group discussions with older pupils, and inviting younger pupils to write 'compositions' on a range of topics. Interviews were conducted with teachers, school governors and other adults in the community. Locke and te Lintelo (2012) explore how young Zambians make sense of delayed or partial transitions to adulthood, with the objective of contributing to conceptualizations of youth transitions. Sixty interviews focused on life trajectories from birth and particularly the social dynamics behind education, aspirations and employment. Using purposive sampling, a selection of young women and men, who were formally and informally employed, self-employed (including some 'subsistence' farmers) and unemployed, was made. Sixteen key informant interviews were also conducted with representatives of youth organizations and donors, and with researchers.

Kristensen and Birch-Thomsen (2013) study the economic activities of rural youth at two locations in the northern province of Zambia and two in eastern Uganda, including the barriers to, and opportunities for, successful business creation. They analyse mobility patterns and what motivates young people to migrate or stay put, and the assistance that is 'visible' to rural youth and whether it matches their needs. The methodology combined a household questionnaire survey (N=164 households, including 387 young people between 15–36 years), FDGs with young people and key informant interviews. Berckmoes and White (2014) study the perceptions of rural youth in Burundi, focusing on livelihoods challenges and transitions, and strategies in relation to education, marriage and migration. The gendering of youth transitions is a principle concern. Interviews and group discussions were conducted with female and male youth, community leaders and non-governmental organization (NGO) facilitators, and a survey was conducted in three villages (N=161). Reynolds-Whyte and Acio (2017) focus on the challenges that rural youth face in accessing land in the Acholi region in northern Uganda, a region affected by over 20 years of armed conflict. A gendered analysis of how social institutions and gerontocracy complicate young people's access to land is developed from 46 individual interviews conducted in one sub-county.

All of the primarily qualitative studies, apart from Levy and Hossain (2014), are designed with an explicit focus on youth. They directly involve young people through interviews and other qualitative methods. Often, the interactions with youth are complemented by interviews with parents, community leaders, local officials or NGO representatives, and sometimes with surveys. However, it is also striking that both the large- and small-scale studies rely mostly on conventional qualitative methods including individual interviews and FDGs. This is despite the recognition that participatory research methods are particularly appropriate with young people (Johnson *et al.*, 2017) as they can provide greater scope for self-expression. One rare example is Daum's (2019) recent study on the aspirations of young people in Zambia's eastern province: he asked 53 young people to draw 'the farm they want' and then complemented this exercise with in-depth interviews. While some studies include both qualitative and survey methods (Ahaibwe *et al.*, 2013; Kristensen and Birch-Thomsen, 2013; Andriamanalina *et al.*, 2014; Berckmoes and White, 2014; Yeboah *et al.*, 2017; Andersson Djurfeldt *et al.*, 2019), survey data are used only to provide simple descriptive statistics.

Most of the studies recognize that while youth constitute a diverse social category, they are often treated as if they are homogenous. However, while an acknowledgement of the importance of diversity informs most sampling strategies, and gender-balanced samples are the norm, the depth of any gender analysis varies and consideration of other aspects of social difference is limited. Few of these studies analyse gendered experiences in as much depth as Ansell (2004), Berckmoes and White (2014), and Reynolds-Whyte and Acio (2017). In terms of other aspects of social diversity, Yeboah *et al.* (2020) is an exception in that migrant and local youth, including youth with different educational backgrounds, were sampled. While the majority of young people in rural areas are likely to belong to a low-income category, none of the studies samples both relatively better-off and poorer youth. Only Leavy and Hossain (2014) present some findings for better-off youth, while Locke and te Lintelo (2012) discuss the life trajectories of relatively poorer youth separately. Andersson Djurfeldt *et al.* (2019) discuss youth access to land and agricultural opportunities vis-à-vis their positionality within households.

Across both large- and small-scale studies there is considerable variation in how much contextual detail is provided and how it is integrated into the analysis. Single location studies including Reynolds-Whyte and Acio (2017) in northern Uganda, and Berckmoes and White (2014) in Burundi, are more likely to discuss context in greater depth. Multi-country studies by Leavy and Hossain (2014), and Petesch and Rodríguez Caillava (2012), give little attention to the specifics of context and, for example, how these might matter for outcome variables such as work, education or migration. The study by Kristensen and Birch-Thomsen (2013) is an exception in interpreting strategies of young people in Zambia and Uganda as a function of their location (primarily in terms of access to farmland). It is interesting to note that most multi-sited studies highlight the shared experiences of young people across very different sites and contexts, rather than analysing how variation in economic opportunities result in different attitudes and strategies. For example, despite working in rural, industrial and urban locations, Locke and te Lintelo (2012) provide no analysis of the differential influence of context on youth transitions. The emphasis is rather on commonalities across the three sites, with one exception being the observation that while rural youth face more structural difficulties in attaining an education and finding work, at least they are farming while 'waiting' for a complete transition (p.788).

Mixed methods studies

Several of the studies in our selection explicitly feature a mix of qualitative and quantitative analyses (Ahaibwe *et al.*, 2013; Kristensen and Birch-Thomsen, 2013; Andriamanalina *et al.*, 2014; Yeboah *et al.*, 2017; Andersson Djurfeldt *et al.*, 2019). In practice, however, these studies tend to emphasize one or the other in their analysis. For example, some primarily offer a qualitative study with a small survey for descriptive statistics added on (Kristensen and Birch-Thomsen, 2013). Conversely, Ahaibwe *et al.* (2013) and Andriamanalina *et al.* (2014) offer descriptive analyses of large quantitative data sets, which are bolstered with FGDs to help inform interpretation of results. Yeboah *et al.* (2017) is different in that the primary study methodology – Q Methodology – is inherently mixed: it organizes data collection on subjective views in such a way that quantitative analyses are facilitated. However, it shares many of the other characteristics of qualitative studies, including small sample size, emphasis on description and elicitation of subjective viewpoints. Andersson Djurfeldt *et al.*'s (2019) study of youth farming in Mkushi, Zambia, comes closest to methods integration. It uses survey data (albeit on a relatively small sample of 268 households) to document trends in agricultural intensification for farm managers of different ages, along with qualitative interviews to describe more detailed patterns, subjective constraints and aspirations for agricultural participation by young people occupying different household positions.

While not included in our list of 39 studies, the work of the Young Lives programme in Ethiopia deserves mention (Chuta and Morrow, 2015; Favara, 2017; Favara *et al.*, 2018; Pankhurst *et al.*, 2018). Young Lives was a longitudinal comparative research programme, that in Ethiopia undertook a 15-year study, following two cohorts of children in 20 sites selected from five regions. The research followed one cohort from infancy to adolescence (aged 1–15), and the second from early childhood to early adulthood (aged 8–22). Combining both quantitative and qualitative data, the studies provide a unique perspective on the effects of poverty on children's lives and transitions. We do not include these studies in our review for two reasons: first, they did not arise through our literature selection process, as described earlier in this chapter; second, with specific respect to the Young Lives programme outputs, the emphasis has been largely on children, rather than on young adults.

The dearth of mixed methods studies deserves comment. On the one hand, mixed methods studies are on the rise in the applied social sciences (Timans *et al.*, 2019). On the other hand, fully integrated mixed methods studies continue to prove challenging (Roelen and Camfield, 2015; Palinkas *et al.*, 2019). Although some big picture assessments are sometimes referred to in framing qualitative studies (e.g. Filmer and Fox, 2014; Losch, 2016), the reverse is seldom true. Furthermore, the qualitative and quantitative studies generally have very little overlap in terms of cited literature (this is particularly evident with migration studies).

Approaches and Methods That Underpin This Book

General approach

The research that is at the heart of this book was conceived with an ambition to use a combination of methods that would allow us to address some of the limitations of the existing empirical studies noted above. In particular, we wanted to design multi-case study research that investigates contextual differences and how these impact on young people's choices, and thus how different economic structures and opportunities shape their actions and aspirations. We also wanted to overcome the limitation of some survey research that is not targeting young people themselves, by designing qualitative methods that directly involve young women and men. Thus, we proposed a mixed methods approach that would integrate analysis of existing country-level LSMS–ISA survey data, and particularly the underexploited panel data available for some countries, with qualitative research in four countries, two in East Africa and two in West Africa. For the qualitative field work we were particularly interested in methods that would privilege the histories, experiences and perspectives of a diverse selection of young people, living in a variety of rural contexts. By including a diversity of young people, as we will explain below, the study aims to overcome shortcomings in existing studies that tend to disaggregate for gender but do not analyse other forms of social difference.

We planned to use a sequential mixed methods design (Creswell *et al.*, 2003), in which quantitative analysis is used to help orient the qualitative research. One particular aim was to use the LSMS–ISA data to help identify 'hot spots' and 'cold spots' of agricultural commercialization (Yeboah *et al.*, 2020), which could then be sampled through field work. Here, the interest was in whether different commercialization contexts affect young people's economic activities and imagined futures. Another aim was to use the field work to explore key relationships identified through the quantitative analysis. Our hope was that this would allow the research to be 'mixed methods' by design, as opposed to simply combining the quantitative and qualitative results at the write-up stage.

In the end we fell short of this ambition. Logistical issues meant that for some countries the quantitative analysis was not sufficiently progressed by the time field sites needed to be identified. These issues also limited our ability to explore in the field, relationships that had been previously identified. In a further complication, political unrest in Burkina Faso meant that we were unable to do field work there; while a decision was eventually made to work in Côte d'Ivoire instead, this country is not involved in the LSMS–ISA data collection initiative.

Quantitative data

The quantitative data used in this book come from a core set of nationally representative household surveys from Burkina Faso, Ethiopia, Niger, Nigeria, Tanzania, Uganda and Zambia. With the exception of Zambia, these data come from the LSMS–ISA project (Carletto and Gourlay, 2019); data for Zambia are from the Rural Agricultural Livelihood Survey (RALS) conducted by Indaba Agricultural Policy Research

Institute (IAPRI), Michigan State University (MSU) and the Zambian Central Statistical Office (CSO, 2012).

The stated objective of all of these nationally representative panel surveys is to improve the understanding of the links between agriculture, socio-economic status, and non-farm income-generating activities. Such surveys have historically focused on the collection of household-level as well as individual-disaggregated data in the areas of demographics, education, agriculture, health and wage employment. One of the features of such surveys is that they collect data on a wide range of activities of relevance to the rural economy, including agricultural production, land ownership, management and control, input use, non-farm employment, and other income and durable asset ownership. Unlike earlier generations of household surveys based on small samples, the large coverage of these data sets allows some disaggregation by geography, household type and other factors.

LSMS–ISA data in particular have provided the basis for much recent analysis of rural livelihoods and associated welfare outcomes (Jones *et al.*, 2014; Aguilar *et al.*, 2015; Coromaldi *et al.*, 2015; Doss *et al.*, 2015; Karamba and Winters, 2015; Kidoido and Korir, 2015; McCarthy and Kilic, 2015; Oseni *et al.*, 2015; Slavchevska, 2015). For example, Sheahan and Barrett (2017) employ LSMS–ISA data to update the understanding of agricultural input use in SSA. In a similar spirit, Palacios-Lopez *et al.* (2017) use LSMS data for six SSA countries to quantify the share of agricultural labour supplied by women.

Depending on the variable, the household surveys used in our analyses collect information at either the household or individual level. Thus, our analysis includes a combination of some information about the activities of the young people in the samples (e.g. labour allocation, migration decisions), and some information about the households with which they are associated (e.g. income portfolios and land rental market participation). This is less than ideal, but reflects the fact that the surveys were not designed with specific analyses of young people's economic activities in mind. Another related limitation is that for at least some households, all responses were provided by the household head, even if the questions related to the activities of a young person (who may or may not have been present).

Given our focus on youth and the rural economy, we exclude households in major urban areas (e.g. Dar es Salaam in Tanzania, Kampala in Uganda, Addis Ababa in Ethiopia, Niamey in Niger, Lagos in Nigeria) and those in areas with estimated population densities of 1000 or more persons/km^2. For those countries with more than one production season, we focus on the main farming season. To produce nationally representative statistics, we use sampling probability weights.

We are particularly interested in the early stages of livelihood building, and as analysis of these data demand an age-delimited definition of youth, we classify individuals in the age range of 15–24 as 'youth'.[v] We deviate from this definition at various points, however. For example, in constructing household-level summary statistics, we find that the sample of household heads in this age range is very small and unlikely to generate reliable statistics (let alone enable further subdivision by geographical or other variables).

Our analysis of these survey data focused on how youth labour allocations could be compared to those of older people (Abay *et al.*, 2020), whether young people farm differently (see Chapter 6, this volume), young people's mobility and migration, and their subjective well-being (Chamberlin and Sumberg, 2020).

Qualitative data

The qualitative component of the research was conducted in Uganda, Ethiopia, Nigeria and Côte d'Ivoire. With the exception of Côte d'Ivoire, these countries were originally selected because they had completed multiple rounds of LSMS–ISA data collection and we wanted both East and West African countries represented. We also wanted to include countries with both economically vibrant and disadvantaged regions that are clearly distinguishable. The design sought to balance country-level representation and youth diversity with the collection of rich, in-depth data. This was done by selecting four different geographical locations per country, devising a sampling strategy to identify participants from different backgrounds, and employing a combination of different qualitative research methods.

Selection of the four sites in each country involved consideration of levels of agricultural commercialization and economic vibrancy. In commercialization hot spots, local economies are dynamic and while farming remains important, there are opportunities for farming commercial crops and for non-farm employment and self-employment (Yeboah *et al.*, 2020, p.143). In cold spots, on the other hand, the local economy is dominated by small farm production based primarily on family labour, while the farm-service economy and the non-farm economy are limited. In each country, two sites were located in hot spot areas with relatively better access to urban markets and road networks, and two were in cold spots, which were relatively more remote with limited infrastructure and low levels of agricultural commercialization and economic vibrancy. Higher levels of commercialization were assumed to influence both actual opportunities available to young people, and how they imagined their futures. In Uganda, for instance, two sites were located in Central region, only 60 km away from the capital and where cash crops are an important part of the agricultural economy. The other sites were located in the Acholi region, the overall economic development of which has been hampered by a conflict that ended in 2006. All 16 study sites are shown in Table 2.1. The analysis for Korhogo, Côte d'Ivoire, was incomplete at the time of publication, so is included only where data are available.

Within each site, participants were identified using purposive sampling to reflect diversity. Table 2.2 presents an overview of the research instruments and sampling strategy. Apart from seeking to achieve gender balance in the selection of participants, all research exercises were designed to enable gender analysis, with questions related to how family dynamics, work and educational opportunities are gendered.

The qualitative research was designed to capture data on key themes in the lives of rural youth: education, economic activity, family and peers, migration, and imagined futures. The idea was that this would allow us to explore connections between issues that are usually dealt with separately in qualitative studies. The design allowed for a sample size that would help mitigate issues of validity and generalizability, which are associated with small qualitative studies. Overall, the qualitative work included 64 FGDs, 416 interviews with young people and 92 interviews with adults. In order to be able to analyse diversity,

Table 2.1. Identification of sites where qualitative research was undertaken, indicating stylized contrast characterization of rural hot spots and cold spots (see Chapter 1, Table 1.2, this volume).

Site	Uganda	Ethiopia	Nigeria	Côte d'Ivoire
1	Awach (Gulu District, Northern Region) [relatively hot]	Wondo Genet (Sidama Zone, Southern Nations, Nationalities and Peoples Region) [relatively hot]	Oba Oke (Olorunda LGA, Osun State) [relatively hot]	Soubré (Bas-Sassandra District) [relatively hot]
2	Luweero (Luweero District, Central Region) [relatively hot]	Bora (East Shewa Zone, Oromia Region) [relatively hot]	Igbariam Eziafor (Anambra East LGA, Anambra State) [relatively hot]	Odienné (Denguélé District) [relatively hot]
3	Mucwini (Kitgum District, Northern Region) [relatively cold]	Jabi Tehnan (West Gojam Zone, Amhara Region) [relatively cold]	Idi Amu and Igbokiti (Olorunda and Egbedore LGAs, Osun State) [relatively cold]	Belle-Ville, Daloa (Sassandra-Marahoué District) [relatively cold]
4	Butuntumula (Luweero District, Central Region) [relatively cold]	Kuyu (North Shewa Zone, Oromia Region) [relatively cold]	Umumbo (Ayamelum LGA, Anambra State) [relatively cold]	Korhogo (Savanes District) [relatively cold]

Notes: LGA = local government area.

Table 2.2. Data collection instruments and sampling used for qualitative research.

Research exercise	Objective	Sampling frame in each research site	Number completed
Focus group discussion (FDG)	General overview of economic activities common for the area – specifically those undertaken by youth – and migration patterns.	Two FGDs involving young women and two involving young men from across the site, with participants identified to include the range of available economic activities.	Uganda: 33 Ethiopia: 32 Nigeria: 16 Côte d'Ivoire: 19
Life history interview	Focus on the trajectory that led the interviewee to the economic activities and domestic arrangements s/he is currently involved in. Particular interest in roles of social norms and institutions, migration and imagined futures.	Twelve individuals from across the site (6 females, 6 males), covering the range of engagement with formal education, from little to tertiary/diploma.	Uganda: 70 Ethiopia: 42 Nigeria: 48 Côte d'Ivoire: 50
Livelihood interview	Focus on the current economic activities and domestic arrangements. Particular interest in roles of social norms and institutions, migration and imagined futures.	Four individuals from across the site (2 females, 2 males) covering the range of engagement in economic activities, from farming and livestock to non-farm self-employment and wage work.	Uganda: 58 Ethiopia: 33 Nigeria: 34 Côte d'Ivoire: 24
Photo–voice interview	Focus on the futures that interviewees imagine for themselves, including education, work, domestic arrangements and migration, and factors that may hinder movement toward these imagined futures.	Six individuals, gender balanced from diverse educational backgrounds: a few years of primary school; some years in secondary school; and one who had completed secondary education or further.	Uganda: 20 Ethiopia: 18 Nigeria: 24 Côte d'Ivoire: 25
Adult interview	Provide context and a historical perspective on the site and change in young people's economic activities, and adult perspectives on barriers and opportunities for young people, and their motivations.	Six individuals (3 females, 3 males), who have children who are youth.	Uganda: 47 Ethiopia: 27 Nigeria: 24 Côte d'Ivoire: 23

we purposively selected young people with different educational backgrounds and those involved in different economic activities, while we explicitly designed questions that enabled us to analyse the implications of these different social markers. The methods emphasized individual histories, which helped to avoid seeing young people's lives as static (i.e. a young tailor will not have always been a tailor) and helped us to analyse how they respond to landscapes with more or less opportunity. Finally, using a range of qualitative methodologies enabled triangulation.

Qualitative research in each site started with a number of FGDs for young women and men, aimed at obtaining an overview of young people's economic activities in the area, and migration patterns. Issues raised during the FGDs partly informed the selection of participants for the different interviews.

Life history interviews covered early childhood to the present and focused specifically on events such as changing residence; parents' divorce or death; starting, stopping or resuming school attendance; marriage and children; and shifts in economic activities. For each 'critical junction' reasons for the change were explored, as were the role of family dynamics and gender relations. Purposeful sampling was used to include young women and men of different ages and levels of educational attainment.

Livelihood interviews focused on the factors and trajectories that resulted in the respondents' involvement in their current work. They also explored efforts to expand or diversify their activities and the challenges faced, including the role of education, family and other social networks in addressing these challenges. Purposeful sampling was used to be sure that young people with different levels of access to land and other assets were included, and to select young people involved in common non-farm activities.

Photo–voice interviews involved each participant being given a small digital camera and being asked to take photos that reflected how s/he imagined the future, and what might enable or hinder progress in this direction. The pictures were then printed and used as prompts during a semi-structured interview. As only a relatively small number of these interviews could be completed per site, individuals were purposively selected to achieve a gender balance and a range of education levels.

Adult interviews primarily sought to add historical context and to obtain adult perspectives on the opportunities and barriers for young people, and the role and value of education.

All interviews, with the exception of those from Côte d'Ivoire, were transcribed into English. Each transcript was then read, and the information reorganized and synthesized to follow a common sequence. The resulting 'synthesis profiles' from each site were brought together in site summaries. The synthesis profiles, transcripts and site summaries were then analysed for insights in relation to questions around, e.g. economic activities, education, waithood, land, etc.

Conclusions: Toward a Stronger Empirical Base

This review of the empirical windows on youth engagement in the rural economy highlights the fact that considerable research has been undertaken, from a variety of disciplinary perspectives, motivated by different concerns and questions, and using a range of methodological approaches. That having been said, two points deserve attention.

The first is that little progress has been made in breaking down the barrier between qualitative and quantitative approaches, and as a result, their respective insights are not being integrated into a more nuanced understanding. This significantly diminishes the potential to influence policy and programming (and young people's life chances). One important reason for the persistence of these barriers is how academia is organized (Bryman, 2007). Many journals are rooted in a single discipline and encourage the articulation of contributions in ways that offer few incentives to bring diverse bodies of literature and methodological approaches together. Peer reviewed publications are often limited in size and scope, making it difficult to fully draw out the qualitative and quantitative sides of even a single research question. Qualitative work relies on narrative exposition that requires space; quantitative methods require set-up (conceptual framework, estimation strategy) and interpretation that similarly are not always easily condensed. In effect, constraints imposed by journals make the publication of good integrative studies difficult, even where mixed methods have been employed at the project level.

Nonetheless, lip service is often paid to the need for more integration. For example, the recent flagship report of the International Fund for Agricultural Development (IFAD), *Creating Opportunities for Rural Youth* (IFAD, 2019), notes that:

> Qualitative research is also needed in order to complement quantitative methodologies as a basis for the attainment of a better understanding of the contextual factors that shape youth livelihood outcomes.
>
> (IFAD, 2019, p.272)

Yet, this same report itself draws on background papers that almost exclusively provide analysis of survey data. Of its 21 background papers, only two offer literature reviews that focus on insights from qualitative studies (Trivelli and Morel, 2019; White, 2019), and the only paper that purports to offer a mix of qualitative and quantitative analysis is a political economy study of national policies (Philips and Pereznieto, 2019). While some of the reviews cite some qualitative studies in passing (Doss *et al.*, 2019; Mabiso and Benfica, 2019), their emphasis is on summarizing the results of primarily quantitative analyses. Hence, the research presented in this book was designed to be a mixed methods study in which the qualitative component represented a sizeable part of the programme.

The second observation is that it would appear – although this deserves further analysis – that the primarily qualitative studies do not have much impact on policy formulation. Ironically, it is likely that the richness of these studies, and the sense they may give of 'endless individual cases', hinders their integration into policy processes. On the other hand, is also important to better understand why quantitative analyses are more easily integrated into policy discussions. It is likely that organizational factors play a role: if the analysts and policy makers responsible for promoting (youth) employment are mainly economists and political scientists, they may be primed by their professional training to be more receptive to quantitative analysis and less open to in-depth qualitative data. Such priming also intersects with more practical considerations and other reasons for why quantitative analysis is attractive. Numbers, to begin with, enable measurable comparisons to be made across regions, countries and social groups, in ways that are not straightforward for qualitative summaries. Moreover, well-defined survey sampling frames enable inferential statistics. Thus, estimates of population values of interest (such as mean and variance of household size in a country) are well defined. Second, nationally representative data appeal to external validity concerns, i.e. characteristics and relationships of interest are representative of the broad population of interest, not just particular villages or particular socio-economic contexts. Third, assiduously assembled claims about causal identification – increasingly enabled by randomized control trials and other experimental approaches to impact evaluation – feed into aspirations for evidence-based policy. This means that formal claims about cause-and-effect relationships may be generated and evaluated against formal and well-defined criteria for assessing internal validity. Such claims are generally not credibly made with qualitative approaches.

Such virtues of quantitative analyses can be used to underpin concrete policy recommendations. The fact that much policy making aims at national-level analysis and solutions, only reinforces the drive to decontextualized policy making based on readily digested statistical facts. On the other hand, it is also clear that some, albeit few, primarily qualitative studies have likely influenced dominant narratives about young people, including that they are not able to access sufficient land (e.g. Amanor, 2010; Berckmoes and White, 2014).

Our view is that more nuanced and integrative analyses should be the objective of policy-oriented youth studies in the coming years. The reliance of so much influential analysis on general household surveys should be re-evaluated in the context of our increasing understanding of the limitations of these instruments. In relation to the economic activities of young people, these include: respondent bias (data on individuals from survey responses provided by others); imperfect correspondence between the high variability over time and space in individual economic activities, and the shorthand measures of such in standard instruments; the organization of data collection around the household as the primary unit of analysis; and multi-year gaps in panel observations. For these reasons, the data collected on some kinds of young people's economic activities may be highly attenuated and biased, even opaque.

On the qualitative side, more explicit engagement with internal and, especially, external validity would strengthen the policy relevance of results produced by qualitative approaches. As explained, the qualitative component of our research programme sought to do this by involving a sizeable sample of participants, and by triangulation through using various methods. Clearly, not all qualitative studies seek to influence policy, but for those that do, new approaches to the synthesis and presentation of findings may be needed. Qualitative work that generates explicitly testable hypotheses will stand a much better chance of informing quantitative work that explores causal mechanisms, and thereby influencing policy. There is scope for better comparative research design and analysis that demonstrates the different experiences of youth, presented in such a way that findings can inform policies that aim to target different groups. Qualitative work ought to bring out the importance and implications of diversity and context, without being overwhelmed by them. It can help to uncover where there are important exceptions, conditionalities or nuances to the trends, correlations and causal relationships demonstrated by quantitative research.

Thus, there is certainly some room for incremental improvement on both sides, while investment in specialized and more integrated data collection exercises that better capture the individual, collective and intergenerational dimensions of young people's lives, may well be warranted. Specifically, approaches that better integrate the best of the quantitative and qualitative repertoires are required to more adequately address questions around:

- Critical youth transitions, including household formation, mobility and relocation, moments of going in and out of schooling and/or training, and labour market entry and participation.
- The role of local and national opportunity structures, embeddedness and path dependence in youth transitions, decisions and trajectories, and how these are gendered and influenced by other social identities.
- Young people's perspectives on the kind of work they hope to do, what market opportunities they want and where, and the meaning of 'decent work'. How engagement in economic activities helps achieve other personal and family objectives.
- The role of youth in driving economic, social and political change, and potential linkages between their economic status and political engagement.

Notes

[i] We recognize that a number of review articles have been highly cited (e.g. Leavy and Smith, 2010; Proctor and Lucchesi, 2012), but the focus here is on empirical studies presenting primary data for all or some of their analysis.

[ii] As noted by Blattman *et al.* (2014), 'a large literature assumes that poor, unemployed young men weaken social bonds, reduce civic engagement, and heighten the risk of unrest', although they also note that empirical evidence for these claims is limited. Limited rural opportunities, particularly with respect to land access, have been more clearly linked to rural out-migration, as we discuss in the text.

[iii] Not all papers fit this mould. Blattman *et al.* (2014) use experimental data, collected through a randomized control trial. Elder *et al.* (2015) use a large individual-level survey data set encompassing many countries but rely largely on descriptive statistics.

[iv] Many studies are not nationally representative (e.g. Bezu and Holden, 2014; Blattman *et al.*, 2014; Kosec *et al.*, 2018; Andersson Djurfeldt *et al.*, 2019).

[v] We acknowledge that by most definitions, individuals under 18 years old are considered children. Nevertheless, given their involvement in economic activity, it seemed reasonable to include 15–17-year-olds. See Bhalotra and Tzannatos (2003) for a discussion of the use of LSMS–ISA data to investigate child labour; see also Roelen *et al.* (2020).

References

Abay, K., Asnake, W., Ayalew, H., Chamberlin, J. & Sumberg, J. 2020. Landscapes of opportunity: Patterns of young people's engagement with the rural economy in sub-Saharan Africa. *Journal of Development Studies (DOI: 10.1080/00220388.2020.1808195).*

Adesugba, M. & Mavrotas, G. 2016. *Youth Employment, Agricultural Transformation, and Rural Labor Dynamics in Nigeria. IFPRI Discussion Paper 01579,* Washington, DC, International Food Policy Institute (IFPRI).

AfDB 2016. Africa's youth in the labour market. *In:* AfDB (ed.) *African Development Report 2015. Growth, Poverty and Inequality Nexus: Overcoming Barriers to Sustainable Development,* Abidjan, African Development Bank (AfDB).

AGRA 2015. *Africa Agriculture Status Report 2015: Youth in Agriculture in Sub-Saharan Africa,* Nairobi, Alliance for Green Revolution in Africa (AGRA).

Aguilar, A., Carranza, E., Goldstein, M., Kilic, T. & Oseni, G. 2015. Decomposition of gender differentials in agricultural productivity in Ethiopia. *Agricultural Economics,* 46, 311–334.

Ahaibwe, G., Mbowa, S. & Lwanga, M. M. 2013. *Youth Engagement in Agriculture in Uganda: Challenges and Prospects. Research Series No. 106,* Kampala, Economic Policy Research Centre, Makerere University.

Allen, A., Howard, J., Kondo, M., Jamison, A. & Jayne, T. S. 2016. *Agrifood youth employment and engagement study,* East Lansing, MI, Michigan State University.

Amanor, K. S. 2010. Family values, land sales and agricultural commodification in South-Eastern Ghana. *Africa: Journal of the International African Institute,* 80, 104–125.

Ambler, K., Doss, C., Kieran, C. & Passarelli, S. 2021. He Says, She Says: Spousal Disagreement in Survey Measures of Bargaining Power. *Economic Development and Cultural Change,* 69(2), 765–788 (doi.org/10.1086/703082).

Andersson Djurfeldt, A., Kalindi, A., Lindsjö, K. & Wamulume, M. 2019. Yearning to farm – Youth, agricultural intensification and land in Mkushi, Zambia. *Journal of Rural Studies,* 71, 85–93.

Andriamanalina, B. S., Burnod, P., Rakotomalala, H. & Deschênes, S. 2014. Rural youth, agriculture and access to land: the case of Madagascar. *Conference on Land policy in Africa."The next decade of land policy in Africa: ensuring agricultural development and inclusive growth",* Addis Ababa, 11–14 November 2014.

Ansell, N. 2004. Secondary schooling and rural youth transitions in Lesotho and Zimbabwe. *Youth & Society,* 36, 183–202.

Baah-Boateng, W. 2016. The youth unemployment challenge in Africa: What are the drivers? *Economic and Labour Relations Review,* 27, 413–431.

Bardasi, E., Beegle, K., Dillon, A. & Serneels, P. 2011. Do labor statistics depend on how and to whom the questions are asked? Results from a survey experiment in Tanzania. *The World Bank Economic Review,* 25, 418–447.

Benson, T., Erman, A. & Baulch, B. 2019. Change and rigidity on youth employment patterns in Malawi. *In:* Mueller, V. & Thurlow, J. (eds.) *Youth and Jobs in Rural Africa: Beyond Stylized Facts.* Oxford, UK, Oxford University Press.

Berckmoes, L. & White, B. 2014. Youth, farming and precarity in rural Burundi. *European Journal of Development Research,* 26, 190–203.

Bezu, S. & Holden, S. 2014. Are rural youth in Ethiopia abandoning agriculture? *World Development,* 64, 259–272.

Bhalotra, S. R. & Tzannatos, Z. 2003. *Child Labor: What Have We Learnt? Social Protection Discussion Paper Series No. 0317. ,* Washington, DC, Social Protection Unit, The World Bank.

Blattman, C., Fiala, N. & Martinez, S. 2014. Generating skilled self-employment in developing countries: Experimental evidence from Uganda. *The Quarterly Journal of Economics,* 129, 697–752.

BMZ 2017. *One World – No Hunger: Future of the Rural World International G20 Conference, Berlin, 27–28 April 2017,* Bonn, German Ministry for Economic Cooperation and Development (BMZ).

Bryman, A. 2007. Barriers to integrating quantitative and qualitative research. *Journal of Mixed Methods Research,* 1, 8–22.

Carletto, C. & Gourlay, S. 2019. A thing of the past? Household surveys in a rapidly evolving (agricultural) data landscape: Insights from the LSMS-ISA. *Agricultural Economics,* 50, 51–62.

Chamberlin, J. & Sumberg, J. 2020. Life satisfaction of rural young people in Tanzania (unpublished paper, under review).

Chuta, N. & Morrow, V. 2015. *Youth Trajectories through Work and Marriage in Rural Ethiopia. Young Lives Working Paper 135,* Oxford, UK, Young Lives.

Coromaldi, M., Pallante, G. & Savastano, S. 2015. Adoption of modern varieties, farmers' welfare and crop biodiversity: Evidence from Uganda. *Ecological Economics,* 119, 346–358.

Creswell, J. W., Plano Clark, V. L., Gutmann, M. L. & Hanson, W. E. 2003. Advanced mixed methods research designs. *In:* Tashakkori, A. & Teddlie, C. (eds.) *Handbook of Mixed Methods in Social and Behavioral Research.* Thousand Oaks, CA, Sage.

CSO 2012. *The 2012 Rural Agricultural Livelihoods Survey (RALS) Interviewer's Instruction Manual. Central Statistical Office (Agriculture and Environment Division) in conjunction with the Ministry of Agriculture and Livestock (Policy and Planning Department), commissioned by Michigan State University (MSU) and Indaba Agricultural Policy Research Institute (IAPRI),* Lusaka, Zambia, Central Statistical Office (CSO).

Daum, T. 2019. Of Bulls and Bulbs: Aspirations, Opinions and Perceptions of Rural Youth in Zambia. *Development in Practice,* 29, 882–897.

Diao, X., Fang, P., Magalhaes, E., Pahl, S. & Silver, J. 2019. Cities and rural transformation: A spatial analysis of rural youth livelihoods in Ghana. *In:* Mueller, V. & Thurlow, J. (eds.) *Youth and Jobs in Rural Africa: Beyond Stylized Facts,* Oxford, UK, Oxford University Press.

Doss, C., Heckert, J., Myers, E., Pereira, A. & Quisumbing, A. 2019. *Gender, rural youth and structural transformation: evidence to inform innovative youth programming. IFAD Research Series 44,* Rome, IFAD.

Doss, C., Kieran, C. & Kilic, T. 2017. *Measuring Ownership, Control, and Use of Assets,* Washington, DC, The World Bank.

Doss, C., Kovarik, C., Peterman, A., Quisumbing, A. & Van Den Bold, M. 2015. Gender inequalities in ownership and control of land in Africa: myth and reality. *Agricultural Economics,* 46, 403–434.

Doss, C. R., Catanzarite, Z., Baah-Boateng, W., Swaminathan, H., Deere, C. D., Boakye-Yiadom, L. & Suchitra, J. Y. 2018. Do men and women estimate property values differently? *World Development,* 107, 75–86.

Elder, S., de Haas, H., Principi, M. & Schewel, K. 2015. *Youth and Rural Development: Evidence from 25 School-to-Work Transition Surveys,* Geneva, International Labour Organization (ILO).

Elias, M., Mudege, N., Lopez, D. E., Najjar, D., Kandiwa, V., Luis, J., Yila, J., Tegbaru, A., Ibrahim, G., Badstue, L., Njuguna-Mungai, E. & Bentaibi, A. 2018. Gendered aspirations and occupational trajectories among rural youth: A cross-regional perspective. *Journal of Gender, Agriculture and Food Security,* 3, 82–107.

Favara, M. 2017. Do Dreams Come True? Aspirations and Educational Attainments of Ethiopian Boys and Girls. *Journal of African Economies,* 26, 561–583.

Favara, M., Chang, G. & Sánchez, A. 2018. *No longer children: What do Young Lives children do when they grow up? Transitions to post-secondary education and the labour market, Research Report,* Oxford, UK, Young Lives.

Filmer, D. & Fox, L. 2014. *Youth Employment in Sub-Saharan Africa,* Washington, DC, Agence Française de Développement and The World Bank.

Fox, L. & Thomas, A. 2016. Africa's got work to do: a diagnostic of youth employment challenges in sub-Saharan Africa. *Journal of African Economies,* 25, i16-i36.

Hino, H. & Ranis, G. (eds.) 2014. *Youth and Employment in Sub-Saharan Africa: Working but Poor,* Abingdon & New York, Routledge.

Honwana, A. 2012. *The Time of Youth: Work, Social Change, and Politics in Africa,* Boulder, Kumarian Press.

IFAD 2019. *Creating Opportunities for Rural Youth: 2019 Rural Development Report,* Rome, International Fund for Agricultural Development (IFAD).

Janzen, S. A. 2018. Child labour measurement: Whom should we ask? *International Labour Review,* 15, 169–191.

Johnson, V., Hart, R. & Colwell, J. 2017. Steps for Engaging Young Children in Research: International innovative methods. *In:* Skelton, T., Evans, R. & Holt, L. (eds.) *Geographies of Children and Young People. Part II: Methodological Approaches,* The Hague, Bernhard van Leer Foundation.

Jones, A., Shrinivas, A. & Kerr, R. 2014. Farm production diversity is associated with greater household dietary diversity in Malawi: Findings from nationally representative data. *Food Policy,* 46, 1–12.

Karamba, R. W. & Winters, P. C. 2015. Gender and agricultural productivity: implications of the Farm Input Subsidy Program in Malawi. *Agricultural Economics,* 46, 357–374.

Kidoido, M. & Korir, L. 2015. Do low-income households in Tanzania derive income and nutrition benefits from dairy innovation and dairy production? *Food Security,* 7, 681–692.

Kilic, T. & Moylan, H. 2016. *Methodological Experiment on Measuring Asset Ownership from a Gender Perspective,* Washington, DC, The World Bank.

Kosec, K., Ghebru, H., Holtemeyer, B., Mueller, V. & Schmidt, E. 2018. The effect of land access on youth employment and migration decisions: evidence from rural Ethiopia. *American Journal of Agricultural Economics,* 100, 931–954.

Kristensen, S. B. P. & Birch-Thomsen, T. 2013. Should I stay or should I go? Rural youth employment in Uganda and Zambia. *International Development Planning Review,* 35, 175–201.

Leavy, J. & Hossain, N. 2014. *Who Wants to Farm? Youth Aspirations, Opportunities and Rising Food Prices. IDS Working Paper 439,* Brighton, UK, Institute of Development Studies (IDS).

Leavy, J. & Smith, S. 2010. *Future Farmers: Youth Aspirations, Expectations and Life Choices. FAC Discussion Paper 013,* Brighton, UK, Future Agricultures Consortium.

Locke, C. & te Lintelo, D. 2012. Young Zambians 'waiting' for opportunities and 'working towards' living well: life course and aspiration in youth transitions. *Journal of International Development,* 24, 777–794.

Losch, B. 2016. *Structural Transformation to Boost Youth Labour Demand in Sub-Saharan Africa: The Role of Agriculture, Rural Areas and Territorial Development. Employment Working Paper No. 204,* Geneva, International Labour Organization (ILO).
Mabiso, A. & Benfica, R. 2019. *The Narrative on Rural Youth and Economic Opportunities in Africa: Facts, Myths and Gaps. IFAD Research Series 61,* Rome, International Fund for Agricultural Development (IFAD).
Maiga, E., Christiaensen, L. & Palacios-Lopez, A. 2015. *Are the Youth Exiting Agriculture en Masse?,* Unpublished mimeo.
McCarthy, N. & Kilic, T. 2015. The nexus between gender, collective action for public goods and agriculture: evidence from Malawi. *Agricultural Economics,* 46, 375–402.
Mueller, V. & Lee, H. L. 2019. Can migration be a conduit for transformative youth employment? *In:* Mueller, V. & Thurlow, J. (eds.) *Youth and Jobs in Rural Africa: Beyond Stylized Facts.* Oxford, UK, Oxford University Press.
Mueller, V. & Thurlow, J. (eds.) 2019. *Youth and Jobs in Rural Africa: Beyond Stylized Facts,* Oxford, UK, Oxford University Press.
Okali, C. & Sumberg, J. 2012. Quick money and power: Tomatoes and livelihood building in rural Brong Ahafo, Ghana. *IDS Bulletin,* 43, 44–57.
Oseni, G., Corral, P., Goldstein, M. & Winters, P. 2015. Explaining gender differentials in agricultural production in Nigeria. *Agricultural Economics,* 46, 285–310.
Palacios-Lopez, A., Christiaensen, L. & Kilic, T. 2017. How much of the labor in African agriculture is provided by women? *Food Policy,* 67, 52–63.
Palinkas, L. A., Mendon, S. J. & Hamilton, A. B. 2019. Innovations in mixed methods evaluations. *Annual Review of Public Health,* 40, 423–442.
Pankhurst, A., Woldehanna, T., Araya, M., Tafere, Y., Rossiter, J., Tiumelissan, A. & Birhanu, K. 2018. *Young Lives Ethiopia Lessons from Longitudinal Research with the Children of the Millennium,* Oxford, UK, Young Lives.
Petesch, P. & Rodríguez Caillava, I. 2012. *Voices of Young Villagers in Sub-Saharan Africa. Background paper,* Washington, DC, The World Bank.
Philips, L. & Pereznieto, P. 2019. *Unlocking the Potential of Rural Youth: The Role of Policies and Institutions. IFAD Research Series 57*, Rome, International Fund for Agricultural Development (IFAD).
Porter, G., Hampshire, K., Abane, A., Tanle, A., Esia-Donkoh, K., Amoako-Sakyi, R. O., Agblorti, S. & Owusu, S. A. 2011. Mobility, education and livelihood trajectories for young people in rural Ghana: a gender perspective. *Children's Geographies,* 9, 395–410.
Proctor, F. & Lucchesi, V. 2012. *Small-scale Farming and Youth in an Era of Rapid Rural Change,* London, IIED & HIVOS.
Reynolds-Whyte, S. & Acio, E. 2017. Generations and access to land in post conflict Northern Uganda: 'Youth Have No Voice in Land Matters'. *African Studies Review,* 60, 17–36.
Richards, P. 2005. To fight or to farm? Agrarian dimensions of the Mano River conflicts (Liberia and Sierra Leone). *African Affairs,* 104, 571–590.
Roelen, K., Barnett, I., Johnson, V., Lewin, T., Thorsen, D. & Ton, G. 2020. *Understanding Children's Harmful Work: A Review of the Methodological Landscape. Working Paper 10,* Brighton, UK, Action on Children's Harmful Work in African Agriculture, IDS.
Roelen, K. & Camfield, L. 2015. *Mixed Methods Research in Poverty and Vulnerability: Sharing Ideas and Learning Lessons,* London, Palgrave.
Schmidt, E. & Bekele, F. 2016. *Rural Youth and Employment in Ethiopia. ESSP Working Paper 98,* Washington, DC and Addis Ababa, International Food Policy Institute (IFPRI) and Ethiopian Development Research Institute (EDRI).
Schmidt, E. & Bekele Woldeyes, F. 2019. Rural youth and employment in Ethiopia. *In:* Mueller, V. & Thurlow, J. (eds.) *Youth and Jobs in Rural Africa: Beyond Stylized Facts.* Oxford, UK, Oxford University Press.
Sheahan, M. & Barrett, C. B. 2017. Ten striking facts about agricultural input use in sub-Saharan Africa. *Food Policy,* 67, 12–25.
Shehu, E. & Nilsson, B. 2014. *Informal Employment Among Youth: Evidence from 20 School-to-work Transition Surveys,* Geneva, International Labour Organization (ILO).
Slavchevska, V. 2015. Gender differences in agricultural productivity: the case of Tanzania. *Agricultural Economics,* 46, 335–355.
Sumberg, J., Anyidoho, N. A., Leavy, J., Te Lintelo, D. J. H. & Wellard, K. 2012. Introduction: the young people and agriculture 'problem' in Africa. *IDS Bulletin,* 43, 1–8.

Sumberg, J. & Hunt, S. 2019. Are African Rural Youth Innovative? Claims, Evidence and Implications. *Journal of Rural Studies,* 69, 130–136.

Tadele, G. & Gella, A. A. 2012. 'A last resort and often not an option at all': Farming and young people in Ethiopia. *IDS Bulletin,* 43, 33–34.

The World Bank 2012. *World Development Report 2012: Gender Equality and Development.* Washington, DC, The World Bank.

Thorsen, D. 2013. Weaving in and out of employment and self-employment: young rural migrants in the informal economy of Ouagadougou. *International Development Planning Review,* 35, 203–218.

Timans, R., Wouters, P. & Heilbron, J. 2019. Mixed methods research: what it is and what it could be. *Theory and Society,* 48, 193–216.

Trivelli, C. & Morel, G. 2019. *Rural Youth Inclusion, Empowerment and Participation. IFAD Research Series 45*, Rome, International Fund for Agricultural Development (IFAD).

White, B. 2019. *Rural Youth, Today and Tomorrow. IFAD Research Series 48,* Rome, International Fund for Agricultural Development (IFAD).

Williams, T. & Pompa, C. 2017. *Invisible Lives: Understanding Youth Livelihoods in Ghana and Uganda,* Toronto, MasterCard Foundation.

Yeboah, F. K. & Jayne, T. S. 2018. Africa's evolving employment trends. *Journal of Development Studies,* 54, 803–832.

Yeboah, T., Chigumira, E., John, I., Anyidoho, N. A., Manyong, V., Flynn, J. & Sumberg, J. 2020. Hard work and hazard: Young people and agricultural commercialisation in Africa. *Journal of Rural Studies,* 76, 142–151.

Yeboah, T., Sumberg, J., Flynn, J. & Anyidoho, N. A. 2017. Perspectives on desirable work: Findings from a Q study with students and parents in rural Ghana. *European Journal of Development Research,* 29, 423–440.

Appendix

Table 2A. Selected empirical studies touching on youth engagement in the rural economy in sub-Saharan Africa (referenced in Fig. 2.1).

Empirical study	Number code in Fig. 2.1	Times cited(Google Scholar)[a]
Blattman *et al.* (2014)	1	399
Honwana (2012)	2	371
Filmer and Fox (2014)	3	370
Richards (2005)	4	318
Bezu and Holden (2014)	5	257
Leavy and Hossain (2014)	6	87
Porter *et al.* (2011)	7	62
Tadele and Gella (2012)	8	52
Locke and te Lintelo (2012)	9	52
Shehu and Nilsson (2014)	10	47
Yeboah and Jayne (2018)	11	46
Ahaibwe *et al.* (2013)	12	44
Losch (2016)	13	44
Kosec *et al.* (2018)	14	43
Baah-Boateng (2016)	15	36
Okali and Sumberg (2012)	16	35
Kristensen and Birch-Thomsen (2013)	17	34
Fox and Thomas (2016)	18	30
Thorsen (2013)	19	25
Elder *et al.* (2015)	20	23
Berckmoes and White (2014)	21	20
Hino and Ranis (2014)	22	20
Adesugba and Mavrotas (2016)	23	15

Continued

Table 2A. Continued.

Empirical study	Number code in Fig. 2.1	Times cited(Google Scholar)[a]
Schmidt and Bekele (2016)	24	15
Maiga *et al.* (2015)	25	14
Elias *et al.* (2018)	26	12
Yeboah *et al.* (2017)	27	12
Allen *et al.* (2016)	28	8
Petesch and Rodríguez Caillava (2012)	29	8
Reynolds-Whyte and Acio (2017)	30	5
Andriamanalina *et al.* (2014)	31	2
Andersson Djurfeldt *et al.* (2019)	32	1
Williams and Pompa (2017)	33	1
Daum (2019)	34	0
Mueller and Thurlow (2019)	35	0
Yeboah *et al.* (2020)	36	0
IFAD (2019)	37	N/A
AfDB (2016)	38	N/A
BMZ (2017)	39	N/A

[a]As of 26 June 2020.

3 Are Africa's Rural Youth Abandoning Agriculture?

Justin Flynn and James Sumberg
Institute of Development Studies, Brighton, UK

Introduction

The idea that young people in rural Africa are leaving agriculture is pervasive in both the research and policy literature (White, 2012; AGRA, 2015; Maiga *et al.*, 2015; UN, 2016; Mabiso and Benfica, 2019). Two explanations are commonly offered. The first is that agriculture is seen as 'hard work', 'dirty', 'backward' and 'low status', making it an option of 'last resort' (Juma, 2007; Langevang and Gough, 2012; Tadele and Gella, 2012; Leavy and Hossain, 2014). The second is that even if they want to farm, young people cannot access land (Amanor, 2010; Ahaibwe *et al.*, 2013; Bezu and Holden, 2014; Filmer *et al.*, 2014; Andersson Djurfeldt *et al.*, 2019; Scoones *et al.*, 2019). Nevertheless, recent research on young people's livelihoods reveals that many are involved in agriculture (Beauchemin, 2011; Potts, 2012; de Brauw *et al.*, 2014; Elder *et al.*, 2015; Fox and Thomas, 2016; Betcherman and Khan, 2018; Yeboah and Jayne, 2018; IFAD, 2019; Abay *et al.*, 2020; Yeboah *et al.*, 2020). While some are farming because they have no other options (Tadele and Gella, 2012), others are actively pursuing agricultural work, and through it establishing independent lives for themselves and their families (Yeboah *et al.*, 2020). Some young people are able to use agriculture to accumulate assets like land, housing and means of motorized transportation (Bouzidi *et al.*, 2015a,b; Andersson Djurfeldt *et al.*, 2019; Yeboah *et al.*, 2020).

This chapter presents new evidence concerning young people's engagement with the rural economy, and uses this evidence to ask whether young people are indeed leaving agriculture en masse, and if so, what they are doing instead. The focus is on broad patterns of engagement, and how these are affected by gender, age and other markers of social difference. The discussion is framed by established debates around the emergence and importance of the rural non-farm economy (RNFE), linkages between farm and non-farm activities, and the changing nature of rural livelihoods – all set against a backdrop of structural transformation. While rural-to-urban migration is an important strategy for some young rural Africans, and is to be expected during structural transformation (de Brauw *et al.*, 2014), we do not focus on this, as our interest lies in their engagement with the rural economy (see Chapter 5, this volume).

The chapter is structured as follows. The next section reviews the terminology, concepts and theories that underpin the literature on the rural economy, and identifies key debates addressed by this literature. Following this, we review insights on youth engagement with the rural economy that emerge from recent analyses of nationally representative household surveys. Findings from qualitative work are then presented. The final section discusses these findings and concludes with a discussion of implications for policy.

DOI: 10.1079/9781789245011.0003

Changing Rural Economies

The what and the where of rural economic activity

Africa's rural economies, and the livelihoods that both shape and are shaped by them, are described and analysed using a variety of categories and terms. For example, it is common to divide the rural economy into the farm economy, the non-farm economy, and transfers (e.g. remittances and social protection programmes). Individual economic activities are usually analysed using some combination of the location where the work takes place, the sector, and/or the type of labour market engagement, that together yield categories such as: farm, on-farm, off-farm, non-farm, domestic labour, unpaid labour, self-employment, wage labour, and more recently, entrepreneurship.

Reardon *et al.* (2007) suggest that agricultural employment refers only to work undertaken in the primary production of agricultural products (this includes self-employment and farm wage labour). Thus, farm wage labour is not considered to be part of the RNFE, and should not be treated as equivalent to non-farm wage labour. Further, they use the term 'off-farm agricultural wage labour' to denote that work is undertaken outside the bounds of one's own (or one's family's) farm. Downstream agriculture-related activities, such as food processing are within the RNFE. Along similar lines, Davis *et al.* (2017) include crop and livestock production as two distinct on-farm activities, while agricultural wage labour (i.e. labour undertaken on someone else's farm) is considered an off-farm activity. They then distinguish between these agriculturally oriented activities, and non-farm self-employment and wage employment. In this scheme, transfers and other types of income (e.g. rental income) are also considered as off-farm. It is important to note that in much of the literature, there is little clarity around the operational definitions associated with these terms.

Linkages

The growth of the agricultural sector in a given rural area can contribute to the expansion of the economy beyond agriculture. This is largely attributed to a rise in demand for non-farm goods and services, the provision of raw goods for trade and processing, and to a higher demand for labour leading to rising wages (Reardon *et al.*, 2007; Djoumessi *et al.*, 2020). Linkages between the farm and non-farm economies also work in the other direction, with the RNFE generating growth and employment through its linkages with the agricultural economy (Haggblade and Hazell, 1989; Haggblade *et al.*, 2007; Haggblade *et al.*, 2010).

Haggblade *et al.* (2007) provide a typology of five different kinds of farm–non-farm growth linkages: production, consumption, factor market, productivity and reverse linkages. The terminology of forward and backward linkages is also commonly used:

- *Production* linkages include forward linkages, where the non-farm sector uses agricultural raw materials (e.g. in food processing), and backward linkages, where agriculture uses non-farm sector inputs (e.g. agrochemicals, pumps, etc.).
- *Consumption* linkages come about through expenditure by agricultural households on locally produced goods and services, and expenditure by non-agricultural households on locally produced agricultural products.
- *Factor market* linkages refer to positive effects on the farm sector brought about by investment in the non-farm sector (and vice versa).
- *Productivity* linkages between sectors come about when growth in one sector boosts productivity in the other. For example, lower food costs from increased farm productivity might fuel increases in the productivity of the non-farm sector.
- *Reverse* linkages refer to the benefit that the agricultural sector gains through growth in the non-farm sector. For example, the construction of a local food processing plant (e.g. a juice factory or cannery) can stimulate growth (through increased demand) in the production of local agricultural products.

According to Davis *et al.* (2017, p.154), linkages between agriculture and non-farm activities tend to be weak in rural areas with low levels of economic activity (i.e. there is little activity or growth that can be 'linked'). In these areas, agriculture is generally the main, if not the only engine for rural economic growth (McCullough,

2017). This is particularly the case in much of rural sub-Saharan Africa (SSA), where rural population densities are generally low (particularly compared to many parts of rural Asia) and where agriculture is still the main economic activity (Haggblade *et al.*, 2007). The relative importance of different kinds of linkages will depend on the context and the characteristics of the rural economy. One study of an irrigation scheme in Malaysia found that 80% of non-farm income gains stemmed from consumption linkages, and the remaining 20% from production linkages to agriculture (Haggblade *et al.*, 2007, p.158). This result highlights the potential importance of consumption linkages.

Changing livelihoods

The linkages discussed above help to contextualize the economic conditions under which rural people build and pursue their livelihoods. The majority of young rural Africans are involved in agriculture to some degree (over 90% of those aged 15–16 and around 80% for those 24 and older) (Elder *et al.*, 2015; Fox and Thomas, 2016; Yeboah and Jayne, 2018). IFAD (2019) also finds that young people (aged 15–24) in rural SSA spend close to 70% of their working time in farming (p.238), while those aged 15–34 living in rural and semi-rural areas spend more than 60% of their working time in the agrifood system (i.e. through independent agricultural production and working on farms of others as wage labour, and/or other wage work and self-employment). Finally, Davis *et al.* (2017) suggest that 'regardless of the level of GDP [gross domestic product], agriculture continues to be the distinctive feature of rural livelihoods' (p.155). Nevertheless, there are many rural young people who also work in the RNFE. The process of combining agricultural and non-agricultural activities is commonly referred to as 'livelihood diversification', while a variety of other terms including 'mixed livelihoods', 'portfolio livelihoods' and 'pluri-activity' are also used. Even if SSA's non-farm economy has traditionally been seen as relatively underdeveloped, it is important to remember that the subcontinent's rural residents have combined activities across sectors for decades. Unfortunately, rural people are still often labelled as 'farmers', with the accompanying assumption that they are only engaged in, and totally dependent on, agriculture. The non-farm sector in SSA is now generally seen as undergoing rapid expansion, as an integral part of broader processes of rural transformation (Barrett *et al.*, 2017; Jayne *et al.*, 2018).

The livelihoods diversification literature distinguishes between 'pull' and 'push' factors, with the former including the higher returns individuals and/or households expect to derive from non-farm activities (Reardon *et al.*, 2007). Push factors arise from negative aspects of the agricultural economy that constrain the livelihoods and well-being of agriculture-dependent households. Examples include seasonality and reduced capacity to farm due to short-term shocks (e.g. drought or physical incapacitation), and long-term trends like loss of soil fertility and subdivision of inherited land into increasingly smaller plots. Decisions to diversify taken with no crisis in view (*ex ante*) are sometimes framed as 'risk management strategies', and if successful may lead to an accumulation of assets and additional investment in non-farm and/or on-farm activities (Reardon *et al.*, 2007; Bezu and Holden, 2014; Davis *et al.*, 2017). In contrast, steps to diversify taken in the midst of or following a crisis (*ex post*) are usually classed as 'coping strategies' (Ellis, 1998, 2000; Reardon *et al.*, 2007). Thus, depending on context and household situation, diversification may be driven by either an expanding or a contracting opportunity landscape (Haggblade *et al.*, 2007).

Indeed, among the determinants of diversification, Reardon *et al.* (2007) cite financial and other capitals as allowing a differentiated ability to undertake rural non-farm activities:

> Skill-based and financial barriers to entry do not deter wealthy households, whose members systematically cream off the most lucrative opportunities in the RNFE [. . .] Conversely, asset-poor households remain confined to the low-return segment of the RNFE.
>
> (Reardon *et al.*, 2007, p.134)

Along these lines, the work of Bezu *et al.* (2012) in Ethiopia illustrates how households with a greater capacity to accumulate capital, with more adult labour or better access to credit and savings, are better able to take advantage of high return non-farm activities. In contrast, households which are in low productivity activities, such as subsistence farming, engage in these types of activities as they serve as 'a refuge for the poor'. High barriers to entry keep poor households from undertaking higher return non-farm activities (see also Alobo Loison, 2015; Davis *et al.*, 2017).

Reardon *et al.* (2007) use similar arguments to account for women's more constrained livelihood choices and more limited ability to diversify relative to men. Both of these expectations are borne out in a study of labour by McCullough (2017): across Ethiopia, Malawi, Tanzania and Uganda, women – who tended to have lower levels of education than men – were more involved in agriculture, while men were more involved in the service and industrial sectors. These factors represent a 'dual-burden' for women, limiting both their opportunity space (IFAD, 2019) and their ability to exploit it through diversification.

More broadly, Bryceson and Jamal (1997) and Bryceson (2002) argue that the Structural Adjustment Programmes (SAPs) of the 1980s and 1990s, which removed input subsidies (e.g. for fertilizers, seeds, pesticides), and led to lower yields and reduced agricultural incomes, were significant push factors. In their view the resulting undermining of agrarian livelihoods can be directly linked to ongoing processes of 'depeasantization' and 'deagrarianization'.

It would be expected that processes of livelihood diversification and deagrarianization, if widespread, could be observed through the incomes of rural households. Yet, empirical analyses do not provide a clear picture, with average non-farm income estimated to range from 30% to 80% (Bryceson, 2002; Reardon *et al.*, 2007; Davis *et al.*, 2017). Davis *et al.* (2017) used a nine-country SSA data set to show that the share of agricultural income decreases with every income quintile. In most countries, for the poorest quintile, around 50% of household income comes from on-farm activities (i.e. crop and livestock production only), while this drops to less than 20% for the richest quintile. This supports the suggestion mentioned above that it is the better off who are able to take advantage of more remunerative opportunities in the non-farm economy. Income from agricultural wage employment also decreases with increasing levels of household expenditure. Overall, as total income rises, agricultural income shares decline while non-farm income shares increase, particularly from non-agricultural wages and self-employment.

Structural transformation

Structural transformation refers to the shift in labour from less productive to more productive sectors, thereby generating economic growth (Kilby and Johnston, 1975; Timmer, 1988; McCullough, 2017). Structural transformation is considered by most development economists to be a fundamental feature of economic growth and, in theory, proceeds until productivity levels between the farm and non-farm sectors converge (Diao *et al.*, 2017, p.415). Davis *et al.* (2017) suggest that income diversification is a key characteristic of the transformation process.

For Diao *et al.* (2017, p.426–427), structural transformation is well under way in SSA as evidenced by a 9.3% point fall in labour allocated to the agricultural sector in Ethiopia, Ghana, Kenya, Malawi, Nigeria, Senegal, Tanzania and Zambia, between 2000 and 2010. However, there are clear signs that this transformation is following a different path than that observed in other developing countries and regions. Rodrik (2016a) finds that instead of moving into the manufacturing sector, as in Asia and Europe, labour in SSA is shifting into a service sector that is not particularly productive and that remains largely informal. The growth of cities and rural towns based on services rather than industry has been referred to as 'urbanization without industrialization' (Gollin *et al.*, 2016). For the service sector to be productive, it typically requires high levels of skill (e.g. the information technology (IT) sector), which usually takes an extended period of time to develop, and demands sustained support. As for the industrial sector, Rodrik (2016b) suggests it is peaking much sooner in terms of employment share (at comparable income levels) than seen in the Asian Tiger economies, a phenomenon he terms 'premature deindustralization'.

These limitations notwithstanding, structural transformation is widely seen as a key framework for understanding the evolving opportunities for rural youth employment and livelihoods in rural and urban spaces (Barrett *et al.*, 2017; Diao *et al.*, 2018; Jayne *et al.*, 2018; Mueller *et al.*, 2019).

Method

To gain additional insight into young people's engagement with the rural economy – and particularly the role of agriculture – we draw primarily on the transcripts of 416 individual interviews undertaken in Uganda, Ethiopia, Nigeria and Côte d'Ivoire (see Chapter 2, this volume). The basic approach was to use information on economic activity to divide the research participants in each site into segments representing broadly different patterns of engagement with the rural economy. We approached this not as a statistical clustering exercise, but more in the spirit of broad market segmentation (Yankelovich and Meer, 2006). Nevertheless, the objective was to minimize variation in engagement within the segments, and maximize variation between the segments.

In establishing the segments, the fact of growing maize rather than cowpea, or engaging in petty trade rather than catering was of little importance. Rather, the segments were constructed using a scheme that classifies all paid and unpaid rural work by sector (agriculture, not agriculture); by location (on own farm, not on own farm); and by labour arrangement (unpaid/domestic, self-employment, wage) (Table 3.1). Ultimately, these were reduced to three main patterns of economic activity: agricultural work (including own production, unpaid family farm labour and farm wage work); combined agricultural and non-farm work; and non-farm work (including self-employment and wage work).

Each segment was described by a short narrative (Box 3.1). Segments were compared within sites and across sites within each country. We had a specific interest in comparing segments in sites with different degrees of agricultural commercialization.

Youth in the Rural Economy

Insights from existing quantitative work

Recent analyses based on large-scale household surveys provide some insight into young people's engagement with the rural economy. Abay *et al.* (2020) use national representative household survey data from Uganda, Ethiopia, Tanzania, Zambia, Nigeria and Niger to examine how rural young people allocate their labour. Among

Table 3.1. Simple classification of rural work based on sector and location.

		Location	
Sector	Labour arrangement	On own farm[a]	Not on own farm
Agriculture	Unpaid/domestic	Crop and livestock production Primary processing of agric. commodities	
	Self-employment	Crop and livestock production Primary processing of agric. commodities	Agric. input supply, processing of agric. commodities, produce trade, etc.
	Wage employment		Farm labour Employment with agric. input supplier, produce trader, etc.
Not agriculture	Unpaid/domestic	Caring, cooking, childminding, etc.	Help in household business
	Self-employment	Brick making, mat weaving, etc.	Catering, petty trading, hairdressing, seamstressing, carpentry, transportation, etc.
	Wage employment		Employment in shop, bar/restaurant, teaching, etc.

[a]Own or household's farm.

Box 3.1. Example segment narratives.

Wondo Genet, Ethiopia, Segment A – Combined agricultural and non-farm work

The individuals associated with this segment are all engaged in agriculture. Cultivation of *khat*[a] is common and is the most important source of revenue; it is sometimes grown on irrigated land. Other crops include coffee, *ensete* [false banana], potato and maize, and these are consumed and/or sold. All individuals in this segment also engage in one or more other economic activities such as owning a small shop, helping out in the family shop, working as a barber or a guard. Money from *khat* cultivation was used to start some of these enterprises.

Individuals within this segment live under various household arrangements. Those who are not married, including a few students, mostly live with their parents (a small number live in rented accommodation), while those who are married typically live with their spouse in an independent household. Married individuals mostly live on land belonging to the male spouse, which was given upon marriage. They farm on land made available by the husband's father.

This segment includes women and men aged between 20 and 29. Some are still students; others have only a few years of formal education; still others have completed Grade 10; and a few are pursuing post-secondary education.

This segment narrative is informed by six key interviews, two women and four men.

Soubré, Côte d'Ivoire, Segment B – Non-farm work (only)

Individuals associated with this segment engage with the rural economy through a wide variety of non-farm work, including both self-employment (preparing and selling local food including *placali*[b] and *attiéké*[c]; working as a barber, a beautician, a photographer, a mason and a house painter; operating an *Orange Money* stall) and working for wages (in a bistro; for an insurance broker; doing casual work). The skills needed for jobs like photographer, barber and beautician were mainly gained informally from friends or family.

They do not engage in agricultural activities, either because they do not have access to land, or because they simply do not want to farm (despite having access to land).

Many individuals within this segment take advantage of varying opportunities that arise to build a portfolio of activities. To increase income (e.g. to be able to meet increased parenting obligations) seems to be the primary motivation.

This segment includes women and men, aged between 20 and 30. They are single, married or separated, and live with parents, friends, relatives or spouses.

Most have some secondary-level education, but only one is still in secondary school. The main reasons for leaving school early include a lack of financial support, and for multiple women, early pregnancy.

This segment narrative is informed by eight key interviews: five women and three men.

[a]*Khat* is a shrub native to Ethiopia. Its leaves are chewed as a stimulant.
[b]*Placali* is a staple food made from fermented cassava dough.
[c]*Attiéké* is the Ivorian name for *gari* (i.e. a flour of varying texture made from cassava roots that are grated, squeezed, left to ferment and then dried. It is a major staple food in West Africa).

the economic activities, farming is the most common, with the percentage of 15–24-year-olds that devote some time to farming ranging from 23% in Nigeria to 78% in Uganda. A much smaller share of this age group is involved in either wage employment (from 3% in Nigeria and Niger, to 19% in Tanzania) or non-farm business activities (from 3% in Zambia, to 11% in Tanzania and Ethiopia). In all six countries, a marginally higher share of 25–34-year-olds devote some time to farming compared to the 15–24-year-olds. The percentage of young people involved in wage employment and non-farm business activities increases with proximity to markets, and in both Ethiopia and Zambia, the relative importance of these non-farm activities decreases with remoteness and more slowly in high potential areas.

Yeboah and Jayne (2018) analyse labour market engagement of individual household members across urban and rural areas of Ghana, Nigeria, Rwanda, Tanzania, Uganda and Zambia using data from the Living Standards Measurement Study – Integrated Surveys on

Agriculture (LSMS–ISA) programme of The World Bank and similar households surveys. Using data from 2010–2011 and 2012–2013, and disaggregating 'youth' (aged 15–24) from 'young adults' (aged 25–34), they find that a significant share of youth is economically inactive (ranging from 22.5% in Tanzania to 62.7% in Nigeria). Education is the most commonly cited reason for economic inactivity. Among young adults, economic inactivity is lower (ranging from 1.7% in Rwanda to 25.7% in Nigeria), while females in this age group are more likely to be economically inactive than males. In all six countries, farming provides the largest share of jobs for youth (aged 15–24), ranging from 54.5% in Ghana to 75.5% in Uganda, with the share of non-farm jobs ranging from 18.8% in Uganda to 30.9% in Ghana. The importance of farming is generally less for young adults (aged 25–34), with the share of jobs in farming ranging from 31.8% in Ghana to 64.1% in Rwanda, and non-farm jobs ranging from 28% in Rwanda to 51.7% in Ghana. Males are generally more likely to be employed in the non-farm sector than females.

While useful, these kinds of analyses provide little sense of what young people's economic engagements actually look like on the ground. Using a unique, relatively high frequency data set, Carreras *et al.* (2020) investigate the micro-dynamics of rural young people's work in northern Ghana and eastern Uganda over a 12-month period. The focus is on four dimensions of the work undertaken: its nature, frequency, steadiness and amount. The analysis suggests that the non-domestic work activities of young people are multi-faceted, context and season specific, and highly gendered. The study supports the observation made by others, drawing on different types of data, that the non-domestic work of a significant share of African rural youth is part-time and intermittent, and closely linked to agriculture. Farming, for consumption and sale, is central to the work of many young women and men in both study sites. Many combine farming with other self-employment or wage employment, and those who do tend to work more often. There is little sign of specialization within either their farming or non-farm work. Some young people, particularly women in northern Ghana, work infrequently and for relatively few hours, suggesting that unemployment and underemployment, or a lack of opportunity to work, may well be important. However, these authors caution against assuming that all young people want or are able to work full-time, or all year around, as paid work is only one means of attaining social adulthood (also see Chapter 8, this volume).

Engagement with the rural economy: three segments

Within each of the 16 sites in Uganda, Ethiopia, Nigeria and Côte d'Ivoire, the present study identified three broad patterns (or segments) of engagement with the rural economy. The first, and generally larger segment, includes those who combine own or family farming with some other type(s) of non-farm self-employment and/or wage labour. The second includes those who do not farm, but only engage in non-farm self-employment and/or wage labour. The third segment includes those who only farm: this is a relatively small segment in all sites, ranging from 0% of interviewees in Daloa, Côte d'Ivoire to 35% in Jabi Tehnan, Ethiopia. Across both segments that contain farming, some of this engagement is in the form of unpaid family labour. Across all three segments, many young people are also involved in domestic labour and caring for parents or elderly relatives, siblings' offspring, or their own young families (see also Chapter 9, this volume). In what follows, we provide two examples of individuals associated with each segment.

The first example from the segment that *combines agricultural and non-farm work* is a 28-year-old single man living with his mother and two younger sisters in Awach, northern Uganda. His father died a few years ago, and his two older sisters are married and living elsewhere. Between 1999 and 2005 he lived in various refugee camps due to the war in northern Uganda. He started school in the camps and continued after returning home, where he combined schooling with farming during weekends and holidays. His formal education ended at Primary 5 due to a lack of money for school fees. After this he concentrated on farming. He now pursues a portfolio of economic activities with a distinct seasonal rhythm.

> Farming is very important because it's the source of food and income for our family.

whereas things like running a business can stop at any time, but farming is constant. [. . .] We have over 8 ha of land that we farm. [. . .] Sometimes we sell the farm produce and we also use it for food. [. . .] We all do farming together with our mother. [. . .] I work on the farm more than my sisters, even grazing animals like goats and feeding the pigs. I am the one who does this work more than anybody else at home. Both my older and younger sisters do the same work: farming and domestic work. [. . .] There are activities that I do only during rainy season like farming and burning charcoal. As for brick making, I do this only during the dry season. Other activities like rearing animals continue throughout the year.

The second example from this segment is a 21-year-old married woman living in Kuyu, Oromia, Ethiopia. She lives on her own with her two children; her husband is in prison. A few years ago, he went to do field work with members of the police force, during which a conflict broke out and a policeman was killed. He was suspected of killing the policeman and in 2014 was sentenced to prison. Her formal education went to Grade 10 (she is one of the best educated interviewees at this site). She wanted to continue at a private vocational college in Kuyu, but was unable to pay the fees.

I decide which activities I should do by myself. I was just starting to sell food when my husband was arrested. [. . .] My family are also of low economic status and we do not have many cattle. So, I turned my attention to food selling. [. . .] I support my family by preparing *shiroo* [a stew] and drinks like *farsoo* [local beer]. I do not have land; I own nothing and I do not have any support. I just work and depend on my own labour to make a livelihood for my family. [. . .] I have rented land and cultivated crops as additional work. I rented on contract one *gamsi* for 300 birr for a year. It is very small. What can we do? We just do as our capacity allows. [. . .] From one *gamsi* it is possible to get about four quintals [400 kg] [of wheat]. [. . .] I also engage in *hojii hodhaa* [embroidery]. [. . .] The work I do depends on the season. For example, there is a good market for *araqee* [or *areki*, an alcoholic drink] during the dry season, during the Easter holiday and for wedding ceremonies. I even take it to market to sell. During summer [the rainy season] we engage in farming activities like preparing land, sowing and weeding, and during winter [the dry season] we harvest, collect, store and thresh the crops. I sell food beginning in September when schools start.

A first example from the segment that *only does non-farm work* is a 22-year-old single man who lives with his three younger brothers and two sisters in Igbariam, Anambra State, Nigeria. His father is deceased and his mother lives in another community. Being the eldest, he was able to obtain a National Diploma from a federal polytechnic, but was unable to finish his Higher Diploma due to a lack of money when his father died. His brothers all stopped their education and did not progress higher than secondary school. He wishes he could have finished schooling, so that he 'may get a job any day', but after his father died, 'things changed'.

I learnt how to do the business [of textile trading] as I was an apprentice for some time. [. . .] I [also] work as a labourer, helping bricklayers. [. . .] Usually I am paid four naira per block. If I deliver 400 blocks to a builder at that rate for a day, I will get about 2000 naira, but not always. [. . .] I work as a driver for people in their private vehicles and I earn 1000 or 1200 naira daily. [. . .] Firewood does not sell for very much and we get 500 to 1000 [naira]. [. . .] These jobs are usually available. There is nothing else available to me. No one advised me about these jobs. I got into doing them in order to earn a living to be able to support myself and my younger brothers. [. . .] I always look out for more opportunities. I am never idle; [. . .] I have one paid job or the other.

[Land situation] Well, we have a piece of land but our foster father farms on it and he would not let us use it. We returned to the village 2 years ago and after a family meeting in December he promised to allot some portions to us.

A second example is a 28-year-old married woman living in Luénoufla, a village about 30 km away from Daloa, in south-western Côte d'Ivoire. She lives with her husband (she is his second wife) and their two children, a niece, the first wife and her children, and other relatives of the husband. She attended a Quranic school for 3 years, but stopped when her parents died. She then went to live with her aunt, who would not pay for her schooling. She engages in a mix of activities, but domestic work also takes a lot of her time. She and her co-wife help each other out, including in their business activities.

At the moment, I am selling shoes, along with wrap skirts, earrings and children's clothes and accessories. [. . .] I was selling ox feet and heads

> [when I met my husband]; I was preparing them and selling them. [Have you stopped this activity?] No, I continue, but not that much. [. . .] I do the household chores first. I sweep and when it's my day to cook, I cook. I also do laundry, get the kids ready for school, and when I finish, I cook. The days when I am not cooking, I am at the market selling my wares. [. . .] It is the domestic chores that take most of my time. These chores are the first thing I do when I get up. [. . .] I get up at 4:00 in the morning. [. . .] If I am washing my wares, my co-wife and my niece help me, and when we prepare food, we help each other out as well.
>
> [. . .] Here, apart from cocoa-related activities, nothing else produces money. During the cocoa harvest period there are all sorts of activities here, and thus the young people who are not afraid of work may have money. But after that, it is misery. [When the cocoa activities stop, does it affect your business?] Yes, that's it! The business doesn't work well then.
>
> [Why do you not farm?] Because my parents don't have land. They don't have land, which is why I don't farm.

A first example from the segment that *only farms* (or works on farms) is a 30-year-old married man in Ziévasso, a village about 10 km south of Odienné, in northern Côte d'Ivoire. He lives in a large household with more than 15 members, including his wife and three children, his parents and his four younger brothers. His father farms and makes tools for local gold miners. His mother also farms. He did some farming while going to school, in part to contribute to his school fees, though this made him feel 'overburdened' with work. His formal education ended at lower secondary school ('*troisième*'), because he failed the *Brevet d'Etudes du Premier Cycle* (BEPC) national exam.

> I am a farmer; [. . .] I go to the field daily. [. . .] Today I farm 4 ha. If I have the money, I can do more. [. . .] He [my father] subdivided the land into sections and designated a section for me, as well as sections for each of my brothers. [. . .] I've planted cashew [in my field] and, as the trees grow, little by little, I add food crops underneath. [. . .] I often grow rice, maize, beans. [. . .] This past year I also grew groundnuts [for sauce] [. . .] and potatoes. [. . .] If you have a field of cashew, you can take the money you have made and invest it in another project. Regarding the cereals, you can use them as food. [. . .] I [also] plant small gardens with tomatoes, peppers and okra. When I sell my produce, it helps me [financially].

He describes the importance of his religious and cultural upbringing, which has placed a strong emphasis on duty towards one's parents, and how his sense of duty influences his economic activities.

> According to the customs of our religion, which is Islam, when you reach adulthood [*majorité*] [. . .] you have to take care of your parents. [. . .] [As] my brothers have distanced themselves from the affairs of the household [they are all mining for gold], I have decided to do nothing else but farm work. [. . .] I recognize my responsibility to my parents; they no longer have the strength to work. My duty as a child is to go to the field and do my best to feed them. [. . .] I would have liked to have been a doctor. [. . .] To be honest, I don't regret [not having continued my studies] [. . .] because had I continued, I would not have received the benedictions from my parents for staying close. [. . .] Yes, for me, the benedictions are very important.

A second example is a 20-year-old single woman who lives in Mucwini, northern Uganda, with her parents and four siblings. She went to school up to Primary 5, but stopped due to problems in paying school fees. While at school she farmed in the early morning, and during weekends and holidays. Due to the farming she sometimes arrived late to school.

> My family has 3 acres of land and we have cattle, goats, pigs and chickens. [. . .] The land is owned by my father. He shows me which part of the land to farm and I farm on it for free. I do not own any land. [. . .] I cannot access enough land, so I end up renting land. [. . .] My father helps me farm – he has the oxen for ploughing, and if I ask him to help plough my garden, he does it willingly and for free. I am also in a farming group, where the members help each other in turns. [. . .] I sell my crops at Mucwini trading centre and Kitgum town. [. . .] In addition to farming, I am also rearing chickens, goats, pigs, and I look after the cattle.

One key takeaway from all these respondents is that the majority of young people, with some variation across sites and countries, have some involvement in agriculture. In some sites (i.e. in northern Uganda, Jabi Tehnan woreda in Ethiopia and Odienné region in northern Côte d'Ivoire) farming and/or livestock production is part of the work of nearly all the young people interviewed. Some of their farming is for their own consumption, with any surplus being marketed; some is primarily or solely for sale. While agricultural commercialization is clearly very important, overall, there is little sign of engagement with value chains or quality standards, and the value-added activities that are observed are long standing, such as *gari* production (see Box 3.1). Also, there is little indication that the farming activities of young people are very different from those of their parents, or older people more generally (see Chapter 6, this volume) – indeed, much of the farming of these informants (particularly for those who are younger, single and/or not living independently), is done under, for, or with the support of their elders.

Aside from farming, the self-employment activities in which young people are involved revolve around the common suite of rural economic activities, including petty trade, catering, sewing, hairdressing, artisanal work and the like. Some is seasonal or otherwise ephemeral, and much of this activity can be characterized as generally small scale, requiring little skill, low investment and minimal technology, and presumably, providing low returns. There are few examples of innovative work or 'doing something different', and little sign of any meaningful specialization within the segments. Mobile phones are ubiquitous, and commonly used to contact suppliers and customers, and for mobile banking, but there is little sign of the often-forecast digital agricultural revolution. With the exception of the few young people working in more formal positions, e.g. as teachers, in construction or for non-governmental organizations (NGOs), the wage work is also relatively low skill (such as working as a farm labourer or bar and restaurant work).

The vast majority of young people fall into one of these three segments – farming plus non-farm, non-farm only or farming only – and independent of these segments, many are also involved in domestic and care work. It is also clear that within each segment there are young people who are in very different positions. For example, a low level of education is not always closely linked to engagement in farming, while more education neither guarantees a life without farming, nor a desire not to farm. Similarly, while some interviewees said that compared to young men, it was more difficult for young women to access land, there are a number of examples of young women pursuing their own farming, including the use of land rental markets. In all sites some jobs or activities are seen as more appropriate for women or for men. Nevertheless, the diversity within the segments challenged our own initial assumption that social differences among young people would be critical in influencing broader patterns of engagement with the rural economy.

Depending on the circumstances, parents and other family members might assist the young women and men with their economic activities, but equally, they might not. At the same time, young people must assist with domestic duties and fulfil their moral obligations (i.e. to parents). The broad patterns of economic engagement characterized by the segments do not appear to be particularly sensitive to local economic dynamism, at least as captured by our classification of agricultural commercialization hot and cold spots. Indeed, a stronger driver of engagement in the rural economy looks to be the nature of the region in which a site is situated, rather than the relative dynamism of individual sites.

One distinction that emerges from the segmentation exercise is between young people who are living within a natal household (headed by a parent or older relative), and those who are within an independent household with a spouse or partner (Fig. 3.1). Those who are part of a natal household fall into two subgroups. The first are young, single women and men who are still in school or have recently left school. They may farm some plots for themselves and/or conduct some other independent economic activities, but they also do domestic work, help on the family farm and may contribute to other family enterprises. The second subgroup includes young women who previously left their natal home to live with a spouse or partner, but have returned following divorce or separation (often with one or more children). Again, they may farm some plots for themselves and/or carry out some other independent economic activities, but most often they also help on the family farm or in another family enterprise. We are not suggesting that there are no single people living on their own or with friends, or separated/divorced women and men who live independently. However, among the young people interviewed in the

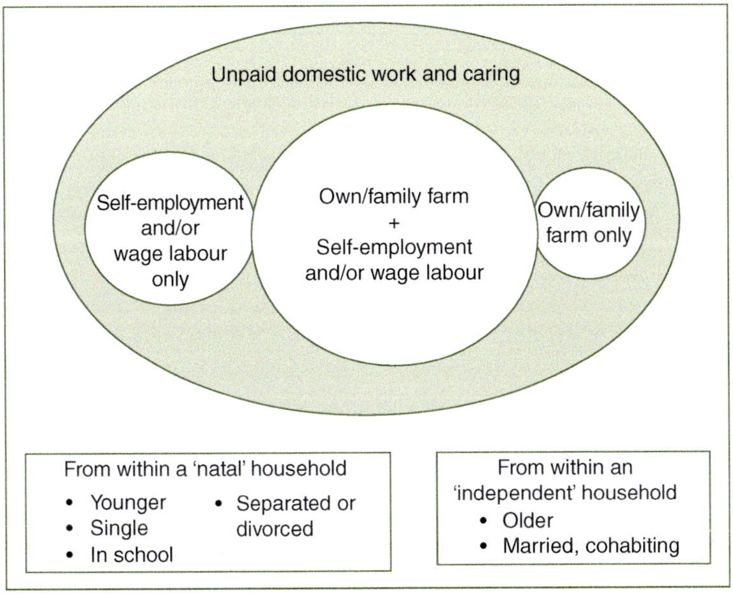

Fig. 3.1. Patterns of engagement with the rural economy. Courtesy of the authors.

course of this research, these were very distinct minorities.

Even if they are pursuing essentially the same economic activities, the key difference between those living in natal and independent households is the degree to which they can exercise autonomy. Young people in natal households are subordinate to the larger unit: they remain, at least in part, under the direct supervision or guidance of elders, and may be wholly or partially dependent on them for land. The work they do on the family farm is likely to be unpaid. In contrast, young people living in their own, independent households farm (and engage in other activities) individually and/or collaboratively, but with considerably more autonomy. Nevertheless, even these households are likely to depend, at least to some degree, on parents or relatives for access to land, and some also benefit from capital and other resources from parents and others.

As developed in greater detail in Chapter 8 (this volume), this latter finding challenges the direct link between engagement in the labour market and the attainment of markers of social adulthood as proposed by Honwana (2012). Rather than the work itself, which is essentially the same regardless of whether they live within their natal home or independently, it is marriage and the platform of an independent household from which the work is undertaken, and the associated autonomy and responsibility, that would appear to drive the transition to social adulthood.

Discussion

Recent quantitative analyses, as well as the analysis of qualitative data from the 16 sites across four countries presented above, strongly suggest that with some variation across sites and contexts, the majority of economically active young people in rural areas continue to maintain some engagement with agriculture. The idea that rural young people have turned their backs on agriculture en masse is clearly not supported by the evidence presented here.

However, three important caveats must be stated. First, few young people are engaged only in farming: the majority of those who farm are also engaging in some other self-employment or wage labour activity. Thus, while there is widespread involvement in farming, in most of the sites it may be quite wrong to conceive of or label these young people simply as 'farmers' (or even 'primarily farmers'). Second, in most cases the engagement of young people with the rural economy does not align at all well with the imaginary of young, highly entrepreneurial, innovative, business and value-chain oriented 'agripreneurs'. A key distinction here is between commercialization on the one hand, and engagement in value chains on the other: the latter includes the former, but agricultural commercialization, in which many young people are involved, does not necessarily mean engagement in value chains. Indeed, much of the farming that young people undertake – on their own or on family plots – is both for own consumption and sale. Even where there are higher levels of commercialization (e.g. in 'hot spots' or where commercial vegetable production is prevalent), there is little direct mention of engagement with value chains, agribusiness firms or certification bodies. This, combined with the fact that farming is one of a number of farm and non-farm economic activities, likely has important implications for if and how young people search out and/or invest in new agricultural technology (Sumberg *et al.*, 2004; see also Chapter 6, this volume). Third, while there is little evidence that rural young people are turning their backs on agriculture, as discussed in Chapter 9 (this volume), few imagine a future of 'digging'. Rather, those who do want to farm envisage more mechanized forms of farming and taking on a managerial role, with increased reliance on hired labour.

Involvement in the non-farm sector is widespread, but most of these activities are small scale, demanding low investment and requiring little skill or technology (see also Flynn and Sumberg, 2017, 2018). They are presumably also low-return activities, suggesting a low-level equilibrium, parallel to the poverty traps described by Barrett and Swallow (2006), and the low-input, low-output farming traps described by Barrett (2008). While this might not be unexpected for young people in the early stages of livelihood building, it is not clear that many of these activities offer viable pathways to secure and/or increased remuneration through upscaling, skill upgrading, innovation or technology investment. The case studies make it clear that many of the young people interviewed were well aware of the limitations of their activities.

The picture that emerges of young people's engagement with the rural economy begs the question as to whether the young people are misreading the rural opportunity landscape, or whether the misreading has been on the part of those promoting the 'rural prosperity' gospel. Rather than a lack of ambition or hard work on the part of young people, it would appear that as far as secure, remunerative or 'decent' work is concerned, the landscape of opportunity in many rural areas remains monochromatic. While the density of opportunities may be greater in commercialization hot spots, it is not immediately obvious that the quality of the opportunities is significantly better.

This analysis calls into question the emphasis on young people's decisions and choices in relation to the economic activities they pursue. For the most part, the initial stages of livelihood building are less about 'career' in some abstract sense, and more about the practicalities of getting established on a path to adulthood. The possible array of economic activity in most rural areas is highly constrained, and this, together with other social and cultural opportunity structures (Roberts, 1968) shapes their engagement with the rural economy. To paraphrase Roberts (1977), by and large, young people simply take what work is available.

Finally, much of the engagement with the rural economy described in this chapter goes beyond the work of individual young people. Rather, whether living in a parental or independent household, their work involves both inter- and intragenerational joint effort, cooperation, exchange and mutual support. To focus solely on the individual and her/his aspirations, choices, decisions and career is to make a fundamental error in conceptualizing both the problem, and its potential solutions.

References

Abay, K., Asnake, W., Ayalew, H., Chamberlin, J. & Sumberg, J. 2020. Landscapes of opportunity: Patterns of young people's engagement with the rural economy in sub-Saharan Africa. *Journal of Development Studies (DOI: 10.1080/00220388.2020.1808195).*

AGRA 2015. *Africa Agriculture Status Report 2015: Youth in Agriculture in Sub-Saharan Africa,* Nairobi, Alliance for Green Revolution in Africa (AGRA).

Ahaibwe, G., Mbowa, S. & Lwanga, M. M. 2013. *Youth Engagement in Agriculture in Uganda: Challenges and Prospects. Research Series No. 106,* Kampala, Economic Policy Research Centre, Makerere University.

Alobo Loison, S. 2015. Rural livelihood diversification in sub-Saharan Africa: A literature review. *The Journal of Development Studies,* 51, 1125–1138.

Amanor, K. S. 2010. Family values, land sales and agricultural commodification in South-Eastern Ghana. *Africa: Journal of the International African Institute,* 80, 104–125.

Andersson Djurfeldt, A., Kalindi, A., Lindsjö, K. & Wamulume, M. 2019. Yearning to farm – Youth, agricultural intensification and land in Mkushi, Zambia. *Journal of Rural Studies,* 71, 85–93.

Barrett, C. B. 2008. Smallholder market participation: Concepts and evidence from eastern and southern Africa. *Food Policy,* 33, 299–317.

Barrett, C. B. & Swallow, B. M. 2006. Fractal poverty traps. *World Development,* 34, 1–15.

Barrett, C. B., Christian, P. & Shiferaw, B. A. 2017. The structural transformation of African agriculture and rural spaces: introduction to a special section†. *Agricultural Economics,* 48, 5–10.

Beauchemin, C. 2011. Rural-urban migration in West Africa: towards a reversal? Migration trends and economic situation in Burkina Faso and Côte d'Ivoire. *Population, Space and Place,* 17, 47–72.

Betcherman, G. & Khan, T. 2018. Jobs for Africa's expanding youth cohort: a stocktaking of employment prospects and policy interventions. *IZA Journal of Development and Migration,* 8, 13.

Bezu, S. & Holden, S. 2014. Are rural youth in Ethiopia abandoning agriculture? *World Development,* 64, 259–272.

Bezu, S., Barrett, C. & Holden, S. T. 2012. Does the nonfarm economy offer pathways for upward mobility? Evidence from a panel data study in Ethiopia. *World Development,* 40, 1634–1646.

Bouzidi, Z., Faysse, N., Kuper, M. & Billaud, J.-P. 2015a. Les projets des jeunes ruraux : des stratégies diversifiées pour accéder au foncier et obtenir l'appui de l'État. *Alternatives Rurales, Hors-Série Jeunes Ruraux,* 13–24.

Bouzidi, Z., Kuper, M., Faysse, N. & Billaud, J.-P. 2015b. Mobiliser des ressources techniques et sociales pour s'installer : stratégies des jeunes ruraux au Maroc. *Cahiers Agricultures,* 24, 420–427.

Bryceson, D. F. 2002. The scramble in Africa: reorienting rural livelihoods. *World Development,* 30, 725–739.

Bryceson, D. F. & Jamal, V. 1997. *Farewell to Farms : De-agrarianisation and Employment in Africa,* Aldershot, UK, Ashgate.

Carreras, M., Sumberg, J. & Saha, A. 2020. Work and rural livelihoods: the micro dynamics of Africa's 'youth employment crisis'. *European Journal of Development Research* (doi.org/10.1057/s41287-020-00310-y).

Davis, B., Di Giuseppe, S. & Zezza, A. 2017. Are African households (not) leaving agriculture? Patterns of households' income sources in rural sub-Saharan Africa. *Food Policy,* 67, 153–174.

De Brauw, A., Mueller, V. & Lee, H. L. 2014. The role of rural-urban migration in the structural transformation of sub-Saharan Africa. *World Development,* 63, 33–42.

Diao, X., Harttgen, K. & Mcmillan, M. 2017. The changing structure of Africa's economies. *The World Bank Economic Review,* 31, 412–433.

Diao, X., Magalhaes, E. & Mcmillan, M. 2018. Understanding the role of rural non-farm enterprises in Africa's economic transformation: Evidence from Tanzania. *The Journal of Development Studies,* 54, 833–855.

Djoumessi, Y. F., Kamdem, C. B. & Ndeffo Nembot, L. 2020. Moving off agrarian societies: Agricultural productivity to facilitate economic transformations and non-agricultural employment growth in sub-Saharan Africa. *Journal of International Development,* 32, 324–341.

Elder, S., De Haas, H., Principi, M. & Schewel, K. 2015. *Youth and Rural Development: Evidence from 25 School-to-Work Transition Surveys,* Geneva, International Labour Organization (ILO).

Ellis, F. 1998. Household strategies and rural livelihood diversification. *Journal of Development Studies,* 35, 1–38.

Ellis, F. 2000. The determinants of rural livelihood diversification in developing countries. *Journal of Agricultural Economics,* 51, 289–302.

Filmer, D., Fox, L., Brooks, K., Goyal, A., Mengistae, T., Premand, P., Ringold, D., Sharma, S. & Zorya, S. 2014. *Youth Employment in Sub-Saharan Africa,* Washington, DC, The World Bank and Agence Française de Développement.

Flynn, J. & Sumberg, J. 2017. Youth savings groups in Africa: they're a family affair. *Enterprise Development and Microfinance,* 28, 147–161.

Flynn, J. & Sumberg, J. 2018. Are savings groups a livelihoods game changer for young people in Africa? *Development in Practice,* 28, 51–64.
Fox, L. & Thomas, A. 2016. Africa's got work to do: a diagnostic of youth employment challenges in sub-Saharan Africa. *Journal of African Economies,* 25, i16-i36.
Gollin, D., Jedwab, R. & Vollrath, D. 2016. Urbanization with and without industrialization. *Journal of Economic Growth,* 21, 35–70.
Haggblade, S. & Hazell, P. 1989. Agricultural technology and farm-nonfarm growth linkages. *Agricultural Economics,* 3, 345–364.
Haggblade, S., Hazell, P. B. R. & Dorosh, P. A. 2007. Sectoral growth linkages between agriculture and the rural nonfarm economy. *In:* Haggblade, S., Hazell, P. & Reardon, T. (eds.) *Transforming the Rural Nonfarm Economy: Opportunities and Threats in the Developing World,* Baltimore, MD, John Hopkins University Press.
Haggblade, S., Hazell, P. & Reardon, T. 2010. The rural non-farm economy: prospects for growth and poverty reduction. *World Development,* 38, 1429–1441.
Honwana, A. 2012. *The Time of Youth: Work, Social Change, and Politics in Africa,* Boulder, CO, Kumarian Press.
Ifad 2019. *Creating Opportunities for Rural Youth: 2019 Rural Development Report,* Rome, International Fund for Agricultural Development (IFAD).
Jayne, T. S., Chamberlin, J. & Benfica, R. 2018. Africa's unfolding economic transformation. *Journal of Development Studies,* 54, 777–787.
Juma, A. 2007. Promoting Livelihood Opportunities for Rural Youth: Some Lessons from Tanzania. *IFAD Governing Council Roundtable 'Generating Remunerative Livelihood Opportunities for Rural Youth'.* Rome, February 2007.
Kilby, P. & Johnston, B. F. 1975. *Agriculture and Structural Transformation: Economic Strategies in Late-developing Countries,* London, Oxford University Press.
Langevang, T. & Gough, K. V. 2012. Diverging pathways: young female employment and entrepreneurship in sub-Saharan Africa. *Geographical Journal,* 178, 242–252.
Leavy, J. & Hossain, N. 2014. *Who Wants to Farm? Youth Aspirations, Opportunities and Rising Food Prices. IDS Working Paper 439,* Brighton, UK, Institute of Development Studies (IDS).
Mabiso, A. & Benfica, R. 2019. *The Narrative on Rural Youth and Economic Opportunities in Africa: Facts, Myths and Gaps. IFAD Research Series 61,* Rome, International Fund for Agricultural Development (IFAD).
Maiga, E., Christiaensen, L. & Palacios-Lopez, A. 2015. *Are the Youth Exiting Agriculture en Masse?,* Unpublished mimeo.
Mccullough, E. B. 2017. Labor productivity and employment gaps in sub-Saharan Africa. *Food Policy,* 67, 133–152.
Mueller, V., Thurlow, J., Rosenbach, G. & Masias, I. 2019. Africa's rural youth in the global context. *In:* Mueller, V. & Thurlow, J. (eds.) *Youth and Jobs in Rural Africa: Beyond Stylized Facts,* Oxford, UK, Oxford University Press.
Potts, D. 2012. Challenging the myths of urban dynamics in sub-Saharan Africa: the evidence from Nigeria. *World Development,* 40, 1382–1393.
Reardon, T., Berdegué, J., Barrett, C. B. & Stamoulis, K. 2007. Household income diversification into rural nonfarm activities. *In:* Haggblade, S., Hazell, P. & Reardon, T. (eds.) *Transforming the Rural Nonfarm Economy: Opportunities and Threats in the Developing World.* Baltimore, MD, John Hopkins University Press.
Roberts, K. 1968. The entry into employment: an approach towards a general theory. *Sociological Review,* 16, 165–184.
Roberts, K. 1977. The social conditions, consequences and limitations of careers guidance. *British Journal of Guidance & Counselling,* 5, 1–9.
Rodrik, D. 2016a. An African growth miracle? *Journal of African Economies,* 27, 10–27.
Rodrik, D. 2016b. Premature deindustrialization. *Journal of Economic Growth,* 21, 1–33.
Scoones, I., Mavedzenge, B. & Murimbarimba, F. 2019. Young people and land in Zimbabwe: livelihood challenges after land reform. *Review of African Political Economy,* 46, 117–134.
Sumberg, J., Gilbert, E. & Blackie, M. 2004. Income diversity, technology choice and agricultural research policy in sub-Saharan Africa. *Development Policy Review,* 22, 131–146.
Tadele, G. & Gella, A. A. 2012. 'A last resort and often not an option at all': Farming and young people in Ethiopia. *IDS Bulletin,* 43, 33–34.
Timmer, C. P. 1988. Chapter 8. The agricultural transformation. *In:* Chenery, H. & Srinivasan, T. N. (eds.) *Handbook of Development Economics, Vol.* 1. Amsterdam and New York, North Holland / Elsevier.

United Nations 2016. *Why are Rural Youth Leaving Farming?* [Online]. United Nations. Office of the Secretary-General's Envoy on Youth. Available: https://www.un.org/youthenvoy/2016/04/why-are-rural-youth-leaving-farming/ [Accessed 21 May 2020].

White, B. 2012. Agriculture and the generation problem: rural youth, employment and the future of farming. *IDS Bulletin,* 43, 9–19.

Yankelovich, D. & Meer, D. 2006. Rediscovering market segmentation. *Harvard Business Review,* 84, 122.

Yeboah, F. K. & Jayne, T. S. 2018. Africa's evolving employment trends. *Journal of Development Studies,* 54, 803–832.

Yeboah, T., Chigumira, E., John, I., Anyidoho, N. A., Manyong, V., Flynn, J. & Sumberg, J. 2020. Hard work and hazard: Young people and agricultural commercialisation in Africa. *Journal of Rural Studies,* 76, 142–151.

4 Young People and Land

Jordan Chamberlin[1], Felix Kwame Yeboah[2] and James Sumberg[3]
[1]*International Maize and Wheat Improvement Center (CIMMYT), Nairobi, Kenya;* [2]*Michigan State University, East Lansing, MI, USA;* [3]*Institute of Development Studies, Brighton, UK*

Introduction

Land is at the centre of much of the discussion of rural young people's economic activities. If starting out in life is the canonical process that defines youth, then many of the decisions and outcomes related to this process will, in one way or another, revolve around land. By starting out in life, we mean transitioning from being a largely dependent member of a household to becoming a more autonomous economic and social agent – although, as is now widely recognized, reality is invariably messier than this simplified transition suggests. In rural areas, the household formation process, an important part of this transition, is most often conditioned by securing land for farming and accommodation. Even though young people are increasingly engaged in non-farm work (Yeboah and Jayne, 2018), and an increasing number may not be doing any 'digging' themselves, few are members of households which are not at least partially dependent on farm production. Considering that sub-Saharan Africa (SSA) is the only region of the world where the rural population, and the number of rural youth in particular, is projected to grow beyond 2050 (Fig. 4.1), access to land will likely remain a critical issue for decades to come.

While the stylized contours of this story have been recognized for some time, its empirical basis has developed rapidly in recent years, driven in large part by the expansion of nationally representative household survey data. Given this volume's inquiry into the patterns of youth engagement in the rural economy, the opportunity to take stock of the empirical knowledge base on youth and land is timely. The purpose of this chapter is threefold: (i) to review available evidence on how young people access and use land; (ii) to clarify linkages between land access and other livelihood decisions and outcomes of interest; and (iii) to identify remaining knowledge gaps and discuss strategies for filling them. The following sections address each of these objectives in turn, after which we summarize the main conclusions.

Young People's Access to Land

Access is increasingly difficult

Land in SSA has traditionally been seen as abundant. In aggregate, the region is sparsely populated, with about 30 persons per km^2 estimated for 2020 (United Nations, 2019a). However, this population is highly spatially concentrated, with 63% of the total rural population found in just 10% of the land area (Jayne *et al.*, 2014). Many parts of rural Africa are experiencing rising population densities, with corresponding increases in land scarcity and shrinking farm sizes (Jayne *et al.*, 2003, 2014, 2016; Knapman *et al.*, 2017;

 Youth and the Rural Economy in Africa: Hard Work and Hazard (ed. J. Sumberg)
DOI: 10.1079/9781789245011.0004

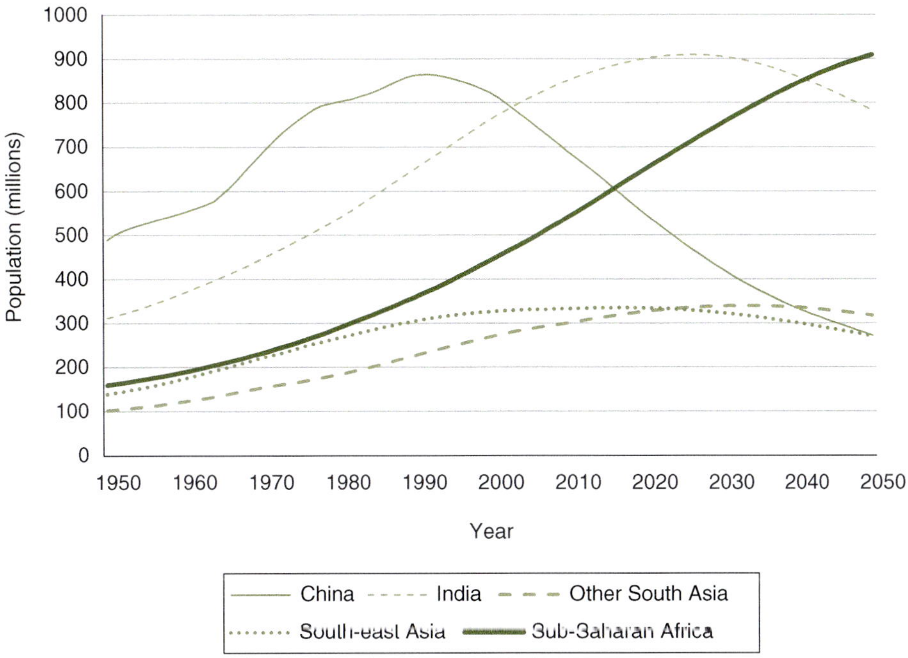

Fig. 4.1. Projected annual rural population (in millions) at mid-year, 1950–2050. From United Nations, 2019b.

Wineman and Jayne, 2018), and increasing perceptions of tenure insecurity (Lawry *et al.*, 2014; Ghebru, 2019). Rising interest in arable land from both local and international investors appears to be exacerbating these pressures in some areas (Jayne *et al.*, 2019). Increasing 'commodification' of land – i.e. when land takes on a monetary value and is exchangeable through market mechanisms – which accompanies rising demand for land by outsiders, is often cited as a major contributor to the difficulties that youth experience in accessing land (Chinsinga and Chasukwa, 2012; White, 2012; Kidido *et al.*, 2017; Asaaga and Hirons, 2019; Kumeh and Omulo, 2019). In systems where land access is increasingly facilitated through market mechanisms (e.g. rental and sales markets), young people may be especially constrained by limited social and capital assets.

Nationally representative surveys indicate clear systematic differences in farm size by age of household head (Fig. 4.2). Although average farm sizes differ widely across countries, reflecting spatial variation in land abundance, within each country, households headed by individuals under 30 years old are associated with smaller farm sizes on average (with the exception of Nigeria, where there is little difference across age groups). Households with heads aged 45 and older generally have the largest farm sizes, often considerably larger than the farm sizes of the youngest heads, although this pattern does not hold in either Ethiopia or Nigeria. Care must be exercised in interpreting this relationship as evidence of young people's increasingly constrained access to land, however. Other factors, such as constrained access to capital, labour, machinery and/or experience in farming, may at least partially explain the observed smaller farm sizes among youth-headed households. Even if access to land were unconstrained, these other factors of production might still limit the farm size of younger households. Thus, while it is likely that young people's access to land is becoming more difficult, the link with farm size is neither direct nor universal.

Changing patterns of access have been documented, with households in densely populated areas having little potential to access additional land through customary land institutions or through inheritance (e.g. Bezu and Holden, 2014; Kosec *et al.*, 2018; Ghebru, 2019). Precise analysis of trends over time is difficult, given limited data availability. However, some indicative

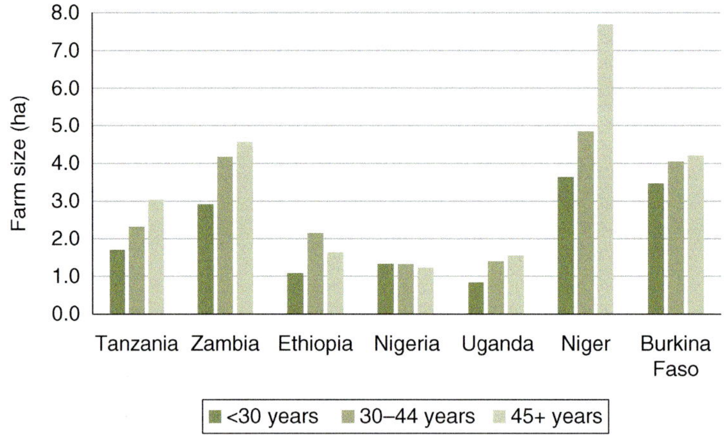

Fig. 4.2. Average farm size (ha) by age of household head. From Living Standards Measurement Study – Integrated Surveys on Agriculture (LSMS–ISA) data for Burkina Faso (2014–2015), Ethiopia (2015–2016), Niger (2014–2015), Nigeria (2015–2016), Tanzania (2012–2013) and Uganda (2011–2012); Rural Agricultural Livelihood Survey (RALS) (CSO, 2012) data for Zambia (2014–2015). Urban households not included in sample. Courtesy of the authors.

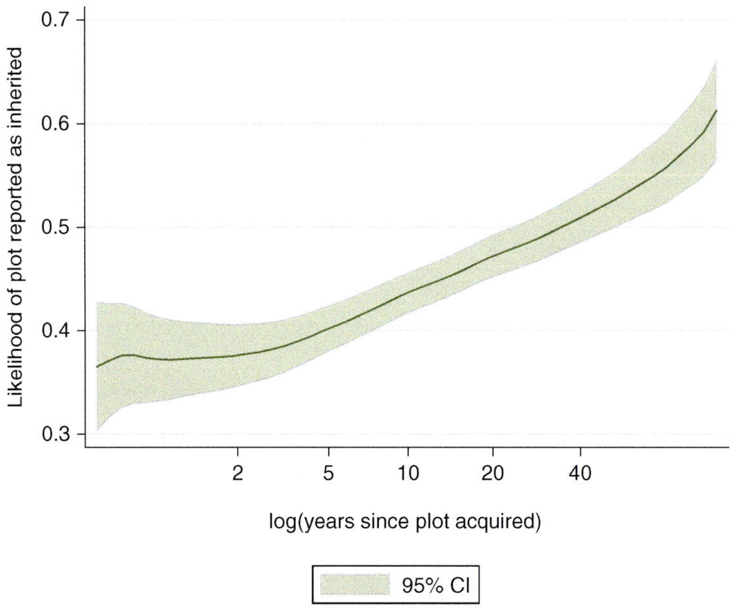

Fig. 4.3. Likelihood of plot being acquired via inheritance, by years since plot acquisition. From Living Standards Measurement Study – Integrated Surveys on Agriculture (LSMS–ISA) data for Tanzania (2014–2015). Courtesy of the authors.

results may be assembled from survey data. Using recent data from Tanzania, Fig. 4.3 shows the likelihood of a plot being reported as having been acquired through inheritance, as a non-parametric function of years since the plot was acquired. More recently acquired plots are much less likely to have been inherited from the preceding generation.

Moreover, buoyed by improvements in health systems and general well-being of the populace, life expectancy in a number of African countries has risen. Indeed, mean lifespan in SSA has increased from 48 years in 1980 to 61 years in 2018, with an estimated growth of 20–42% since 2000 in some countries (The World Bank, 2020). Longer adult lifespans likely mean offspring waiting longer to formally inherit land: while this may not necessarily affect their use of the land in the interim, it might limit their decision making power.

Customary tenure systems seem to be waning in many areas. Sitko and Chamberlin (2016) report that the share of Zambia's land under customary tenure has declined from 94% at independence to at most 54% in 2015. Malawi has experienced similar declines, from 87% at independence to an estimated 60% in 2016 (Anseeuw *et al.*, 2016). Given dwindling stocks of unallocated customary land and the trends in title formalization – which may increase with outsider land acquisitions and the continued development of rural land markets – it seems clear that the ability of young people to access land via customary mechanisms will continue to diminish.

As a consequence of these changes, it is perhaps unsurprising that increasing numbers of rural Africans indicate that surplus land is no longer available in their communities. Even in relatively land abundant Zambia, where the average population density is less than half that of the region as a whole (United Nations, 2019b), fewer than half of rural household heads indicate that unallocated land is available in their area, and that it is possible for them to obtain some of this land for farming (Table 4.1). The trend in perception over time indicates that constraints are intensifying rapidly, and most rapidly among heads younger than 30 years old as well as heads over 45 years. It is possible that pressures on older heads to bequeath land to their children make them more sensitive to local land availability than household heads in the middle age range.

Commodification of land is particularly relevant for young people

The general consensus is that land-related changes are taking place relatively rapidly in many rural areas (Jayne *et al.*, 2019). Rental market participation is increasing (Chamberlin and Ricker-Gilbert, 2016; Muraoka *et al.*, 2018), farm structure is changing, with medium-scale farms showing rapid expansion in many areas (Sitko and Chamberlin, 2015; Jayne *et al.*, 2016), and title conversions are rapidly taking place (Sitko and Chamberlin, 2016). The relative decline in importance of customary institutions as a means of allocating land has been both decried and welcomed as part of a general process of change which has agriculture on a trajectory toward larger-scale, more capital-intensive and commercially oriented production. The commodification of land is an important part of this process. Increasing land scarcity is associated with commodification, and the monetary value of land in rural SSA is increasing in real terms over time (Holden and Bezu, 2016; Wineman and Jayne, 2018; Tione and Holden, 2020), particularly but not only in relatively accessible areas (Tione and Holden, 2020).

Some observers have suggested that the commodification of African farm land will inevitably lead to extractive forms of commercialization that generate few decent employment opportunities, weaken multiplier effects with local rural economies, and exacerbate social differentiation with corrosive effects on community cohesion, welfare and resilience (Cousins, 2007; Amanor, 2012;

Table 4.1. Household heads indicating that unallocated land was available in their area, and that it was possible to obtain some of this unallocated land for farming, Zambia. From Rural Agricultural Livelihood Survey (RALS) (CSO, 2012).

	Age of head			
Survey year	15–29	30–44	45+	Total
2012	47.2%	40.9%	40.3%	41.7%
2015	38.6%	37.1%	35.1%	36.2%
2019	37.1%	36.7%	28.5%	31.7%
Change: 2012–2019	−10.1%	−4.2%	−11.8%	−10.0%

Chitonge *et al.*, 2017a,b, Hall *et al.*, 2017; Yaro *et al.*, 2017; Amanor, 2018). The argument of many of these observers is that commodification will inevitably turn smallholders into landless or nearly landless informal wage workers, with reduced levels of welfare and livelihood security. However, it is important to note that this literature has principally focused on larger-scale commercial investments, and has largely ignored rental and sales transactions between local farmers, which are likely to be the more prevalent modes of commodification in rural Africa. Some of these negative critiques of land commercialization and commodification writ large make claims which are out of sync with their empirical foundations, being frequently very limited in geographic scope, sample size, methodological reproducibility and the types of commodification that are being examined.

The development of rental markets for land is part of the process of commodification, and these markets have increased in importance in recent years. Although still at nascent stages, rural rental market participation rates are rising (e.g. Jin and Jayne, 2013; Chamberlin and Ricker-Gilbert, 2016; Deininger *et al.*, 2017; Ghebru *et al.*, 2018; Ricker-Gilbert and Chamberlin, 2018; Tione and Holden, 2019). This is particularly true for densely populated regions in countries such as Ethiopia, Malawi and Kenya, but it is also evident in more sparsely populated countries such as Zambia (Chamberlin and Ricker-Gilbert, 2016). In land constrained areas, rural land rental markets may be particularly important as a conduit for younger people who want to start farming (Ghebru *et al.*, 2018; Holden and Tilahun, 2018; Kosec *et al.*, 2018; Ricker-Gilbert and Chamberlin, 2018; Yeboah and Jayne, 2018; Yeboah *et al.*, 2019). An important body of evidence suggests that rural land rental markets have important positive effects on equity (by transferring land from land-rich to land-poor households), and overall efficiency (by transferring land to more productive users), and thus contribute to broader positive trends in rural transformation (Holden *et al.*, 2008, 2013; Jin and Jayne, 2013; Chamberlin and Ricker-Gilbert, 2016; Ricker-Gilbert *et al.*, 2019). Offsetting these positive effects, however, there is some evidence that rental rates and other transaction costs are higher for young tenants (Ricker-Gilbert and Chamberlin, 2018), and that rental rates decrease with plot size, making the costs of smaller transactions higher in proportional terms (Abay *et al.*, 2020). Furthermore, in some cases rental markets may be exacerbating access constraints faced by youth in customary areas: White (2012) cites case studies in which community elders were documented as preferring to allocate land to external parties via rental or other arrangements, rather than allowing cultivation rights to local youth. In some cases, this not only constrained youth participation in agriculture, but also fomented social conflict. Thus, while there is good reason to view land markets (particularly rental markets) as representing positive institutional change for increasingly land constrained youth, there is a need for more comprehensive research on the distribution of costs and benefits of these markets, and their potential trade-offs.

Figs 4.4–4.6 summarize land rental market participation patterns at the household level for seven countries in SSA. Several observations stand out. First, there is substantial variation in rental market participation rates across countries: more than a quarter of households in Uganda rent in one or more plots, compared to less than 5% in relatively low-density countries like Niger, Zambia and Burkina Faso. Second, in countries where rental markets are more developed – i.e. Tanzania, Ethiopia and Uganda – relatively younger household heads are significantly more likely to rent in land than older heads. Furthermore, in these same countries, younger households (defined by age of household head) are engaging more intensively with rental markets: they rent in a higher proportion of their farmland and a larger share are renting in 90% or more of their land. This latter statistic implies that many young households would effectively be landless if they did not have access to a rental market.

This inverse relationship between land market participation and age of household head is more visible in Figs 4.7 and 4.8, which use non-parametric regressions to illustrate the relationship and intensity of renting in, measured as the rented in share of farmed land. It is worth noting that it is not the very youngest household heads who are most active in the rental market.

Given that households which are just starting out are likely to have resource constraints which affect their ability to farm, we next examine the likelihood of renting after controlling for

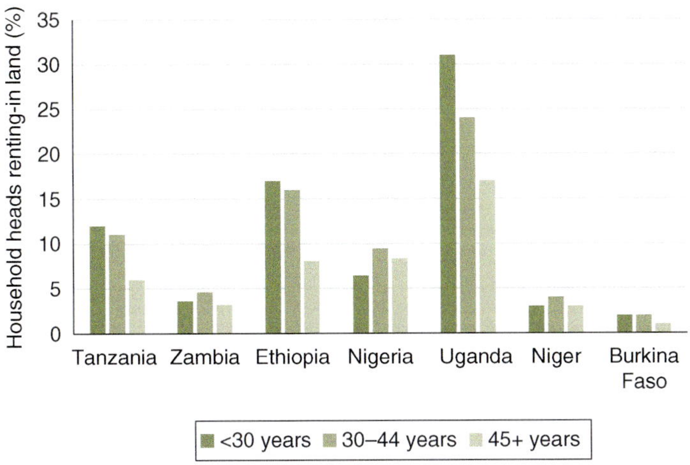

Fig. 4.4. Household participation in land rental market, by age of household head. Analysis of LSMS-ISA and RALS (CSO, 2012) data, courtesy of the authors.

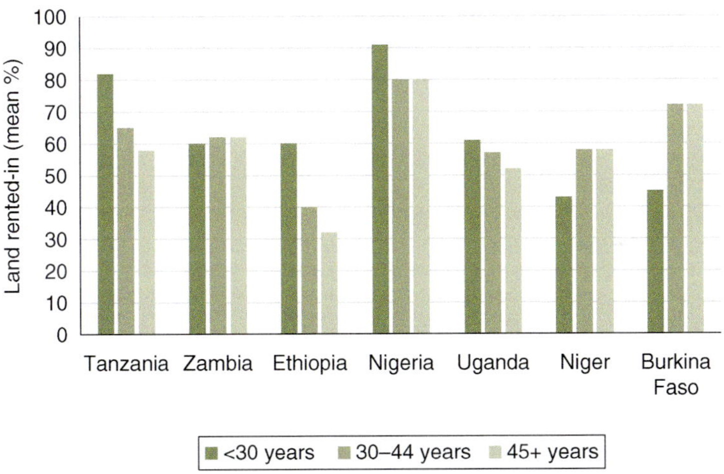

Fig. 4.5. Percentage of land rented in by renting households, by age of household head. Analysis of LSMS-ISA and RALS (CSO, 2012) data, courtesy of the authors.

such factors. Table 4.2 shows the estimated partial effects from Probit models of renting-in land by households in Tanzania, Ethiopia, Uganda, Niger and Nigeria. While there are some differences across the specifications, and in the estimation results, the basic results are very consistent across countries: younger heads are more likely to rent in land, even after controlling for prerental farm size and other observable characteristics.

Using survey data from the Living Standards Measurement Study – Integrated Surveys on Agriculture (LSMS-ISA) programme of The World Bank for Tanzania, from 2008–2009 and 2014–2015 (in which the sample was refreshed), we can compare nationally representative statistics over a relatively short period of time. Surprisingly, these changes are very pronounced and seem particularly so for households with the youngest heads (Table 4.3).

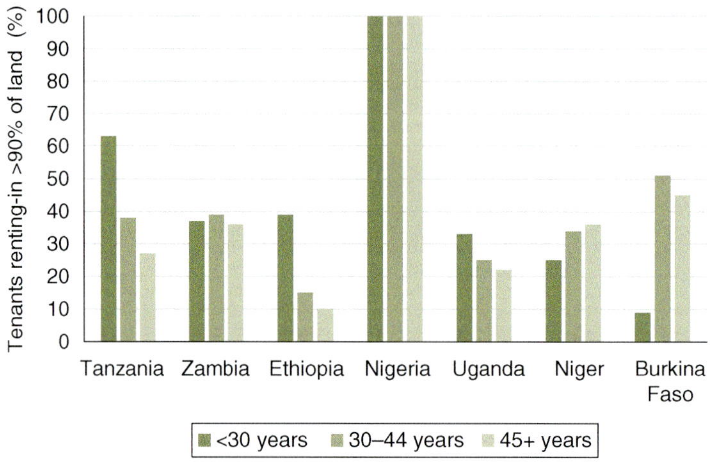

Fig. 4.6. Land renting households that rent in at least 90% of their farmland, by age of household head. Analysis of LSMS-ISA and RALS (CSO, 2012) data, courtesy of the authors.

While overall farm size grew slightly on average over this period, it actually dropped for the youngest households. Median farm size declined over the same period, indicating that the growth in average farm size was driven by relatively few farms at the larger end of the spectrum. Median farm size for the youngest households dropped even more than the median for all households.

Table 4.3 also shows that the general increase in land rental market participation reflects a concentrated use of rental markets by households with the youngest heads. At the same time, the share of total farmland acquired via rental is growing fastest for the youngest households. Growth in the number of plots with title certification[i] is largest for heads aged 30–44, indicating potential age-related constraints (but also likely related to the increased rental share of land in younger hands). Interestingly, while land values are increasing rapidly, this growth has distinct age-related patterns.[ii] Heads under 30 years report faster increases in rental rates than older heads, while the changes in estimated land sales values show the reverse pattern: younger heads report much lower rates of growth. This is consistent with the idea that younger households are more likely to farm marginal lands, in response to relatively higher land access constraints relative to older households.

Paradoxically, given the relatively greater importance of rental markets for young people, there may also be higher costs for them to access such markets. Rural land rental markets are generally still in incipient phases of development, rather thin (i.e. little land on offer and few transactions), and characterized by relatively high transaction costs (Tione and Holden, 2019). In some areas where these markets are more developed, the demand exceeds the willingness of individuals to rent out land, partly because of risk associated with insecure land tenure. Consequently, land prices and rental rates are rising rapidly, particularly in areas of high agroecological potential with favourable access to market (Wineman and Jayne, 2018). In Tanzania, for instance, real land rental rates rose by 5.7% per year between 2009 and 2013, driven largely by improved incentives for farming, urbanization and rising population density, as well as improved tenure security (Wineman and Jayne, 2018). These developments may adversely impact younger renters disproportionately (Holden and Tilahun, 2018; Ricker-Gilbert and Chamberlin, 2018). Tione and Holden (2019) find evidence of large non-linear transaction costs in Malawi; such costs may be particularly hard to negotiate for young potential tenants, suggesting a role for policy action. Ricker-Gilbert and Chamberlin (2018) find evidence in Tanzania that unit rental costs paid by younger tenants were higher

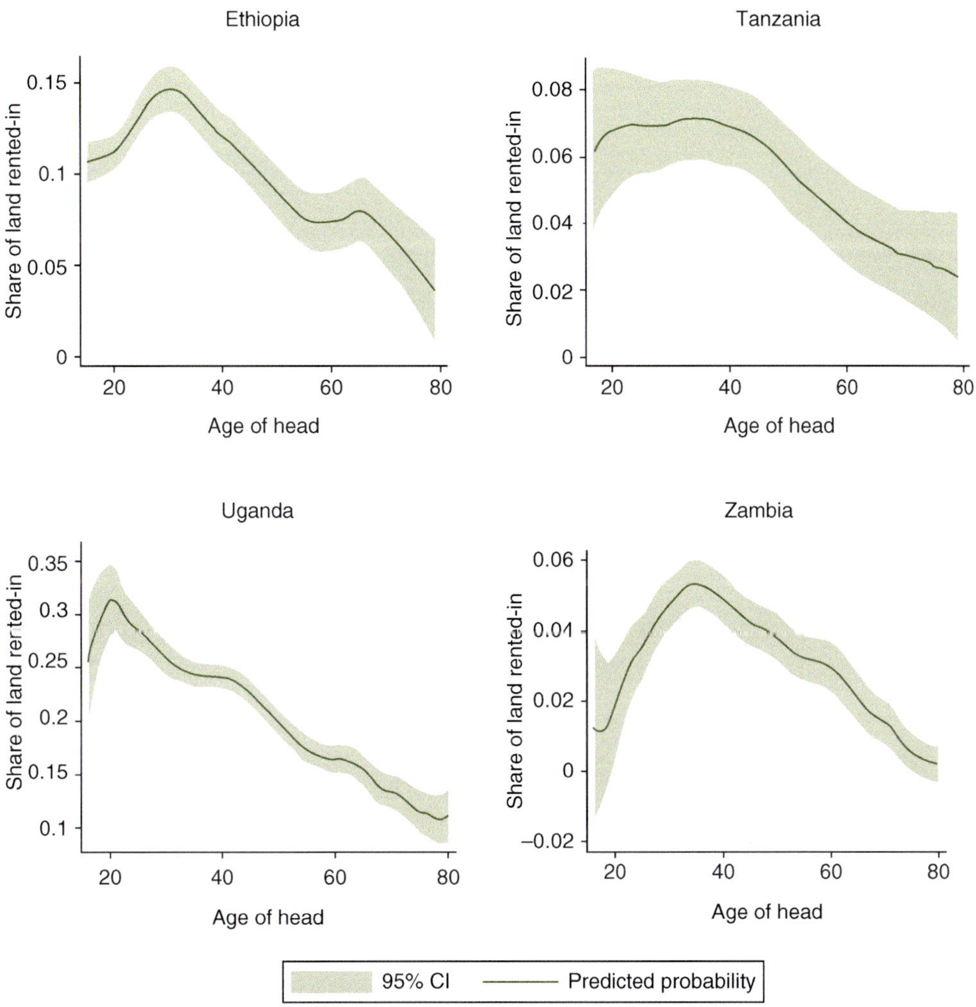

Fig. 4.7. Probability of household renting-in land, by age of household head. Analysis of LSMS-ISA and RALS (CSO, 2012) data, courtesy of the authors.

than those paid by older tenants, suggesting age-related barriers (possibly related to social capital in contexts of weak contractual enforcement).

In counterpoint to these large-scale patterns, qualitative interviews with young people indicate how variable and complex their relationships with evolving land institutions are in different contexts. From among all the 317 interviews conducted across 15 sites in Uganda, Ethiopia, Nigeria and Côte d'Ivoire (see Chapter 2, this volume), 221 young people (70%) reported some own-account farming activity. As shown in Table 4.4, in all sites except Bora and Jabi Tehnan in Ethiopia, family/customary channels or borrowing were the most important ways that young people accessed land. There is considerable variation in the prevalence of 'ownership'[iii] (0% in four sites, rising to 42% in Daloa) and rental (0% in three sites, rising to 38% in Jabi Tehnan in Ethiopia and Oba Oke in Nigeria). Sharecropping was only reported in Ethiopia, where it accounted for between 7% and 31% of cases of reported land use.[iv] Table 4.4 also shows averages for 'relatively hot' and 'relatively cold'

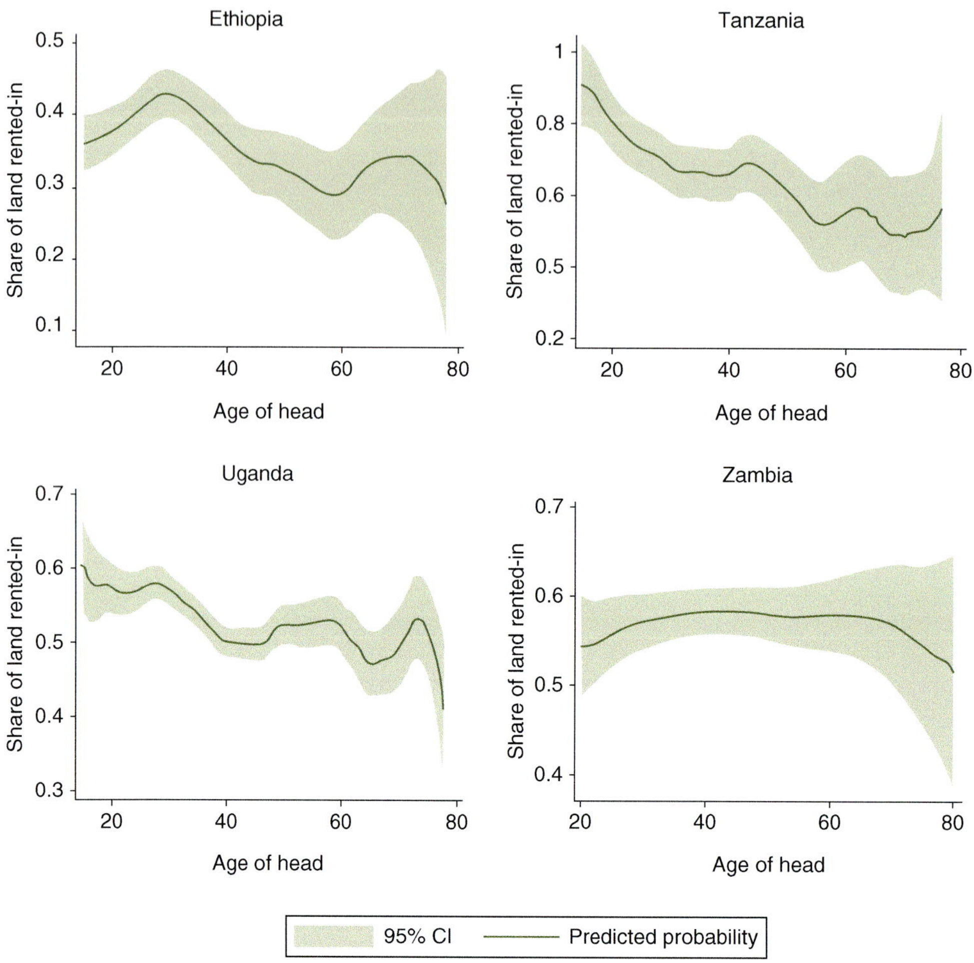

Fig. 4.8. Share of household's land rented-in, by age of household head. Analysis of LSMS-ISA and RALS (CSO, 2012) data, courtesy of the authors.

sites – refering to the level of agricultural commercialization – within countries. Ownership is more important and rental less important in hot spots in both Uganda and Ethiopia (including Wodo Genet where khat production is important), but the opposite trend is seen in Nigeria.

The interviews also picked up a number of cases of young people or their relatives renting out excess land. For some, it was clear that the income derived from renting out land made an important contribution to household welfare.

Both the interviews conducted for this research, and the nationally representative survey data referred to earlier, show that in at least some sites, rental markets play an important role in allowing young people to gain access to land (or to additional land). To date there has been an assumption that young people and others would benefit if landholders brought more of their land into the market. As a general proposition this may be correct, but the magnitude and distribution of these benefits will depend on how rental markets in particular locations actually function. In the remainder of this section, we highlight a number of potential issues identified in the interviews with individual young people: we make no claims about their absolute or relative importance, but they are all plausible and may therefore deserve further attention.

Table 4.2. Determinants of renting-in land (average partial effects from Probit model).

Dependent variable (1 = rents in land)	Tanzania	Ethiopia	Uganda	Niger	Nigeria
Age of household head	−0.0134	−0.0056	−0.0026	−0.0017	−0.001
	(0.000)***	(0.0008)***	(0.0007)***	(0.0007)**	(0.000)**
Prerental farm size (ha)	−0.4356	0.0111	−0.1430***	−0.0160	−0.052
	(0.000)***	(0.007)	(0.0338)	(0.0066)**	(0.019)***
Household size	0.0051	0.0034	0.0111	0.0035	0.0037
	(0.854)	(0.0058)	(0.041)***	(0.0021)	(0.003)
Maximum educational attainment	0.0057	0.0008	0.0036	−0.0009	−0.0004
	(0.721)	(0.0008)	(0.0030)	(0.0027)	(0.001)
Female head = 1	0.0219	−0.1803	−0.0029	−0.0524	0.1644
	(0.929)	(0.0277)***	(0.0268)	(0.0283)*	(0.019)
Number of plots	0.7075		0.0077	0.0081	0.0031
	(0.000)***		(0.0061)	(0.0029)***	(0.002)*
log (Assets)	0.0495				
	(0.005)***				
log (Income per capita)			0.0122	0.0142	0.0056
			(0.0129)	(0.0064)**	(0.065)
Household has ox plough = 1	0.1121	0.084	0.0390	0.0449	
	(0.342)	(0.026)**	(0.0477)	(0.0285)	
Household has tractor = 1	0.0574		0.1806		
	(0.734)		(0.2389)		
Located in rural area = 1	0.0619	0.1085			−0.095
	(0.753)	(0.0386)**			(0.033)***
Kilometres to road	−0.0017	0.0008	−0.0005	0.0010	0.0022
	(0.820)	(0.0009)	(0.0015)	(0.0007)	(0.001)
Kilometres to market	0.0004	−0.0008	−0.0015	−0.0004	0.0004
	(0.943)	(0.0004)**	(0.0009)	(0.0003)	(0.000)*
Elevation	0.0007	0.0001	0.0001	−0.0003	−0.000
	(0.175)	(0.0000)	(0.0001)	(0.0001)**	(0.000)
Slope	−0.0315		−0.0004		−0.003
	(0.237)		(0.0019)		(0.004)
log (Population density)	0.0000	0.0001			0.0018
	(0.974)	(0.0000)			(0.007)
log (Kilometres to night lights)			−0.0091	0.0022	
			(0.0054)*	(0.0035)	
Bimodal rainfall = 1	−0.4316				
	(0.451)				
Mean annual rainfall	0.0015	0.0001	0.0001	−0.0005***	0.000
	(0.262)	(0.0000)**	(0.0001)	(0.0002)	(0.000)***
N	8741		1887	1723	2208

Notes: Tanzania model uses 2008–2009, 2010–2011 and 2012–2013 waves of the Tanzanian Living Standards Measurement Study – Integrated Surveys on Agriculture (LSMS–ISA) data. For Tanzania, Mundlak-Chamberlain controls (time-averages of time-varying regressors) and year dummies are included, but coefficients not reported. Uganda model uses the 2011–2012 wave of the Uganda LSMS–ISA data. Niger model uses the 2014 wave of the Niger LSMS–ISA data. P-values are cluster robust, with significance levels denoted as follows: * = p<0.10, ** = p<0.05, *** = p<0.01.

One important question is whether rental markets are open to all on equal terms. For one 21-year-old woman from Igbariam Eziafor, Nigeria, gender is not a barrier. According to her: 'The plots of land are owned by different people, so you meet a landowner, and settle the price for a plot with him [the landowner] and then he gives it to you'. She seems confident of being able to hire a piece of land to farm: 'If we have the money, we can.' A 19-year-old woman

Table 4.3. Farm size and land market characteristics, by age of household head (HH), Tanzania.

	Age of household head			
Survey year	15–29	30–44	45+	Total
Mean landholding size (ha)				
2009	1.39	2.29	2.61	2.34
2015	1.23	2.32	3.09	2.49
Change: 2009–2015	−12%	1%	18%	7%
Median landholding size (ha)				
2009	0.81	1.21	1.42	1.30
2015	0.61	1.10	1.46	1.17
Change: 2009–2015	−25%	−9%	3%	−9%
% of HHs renting-in land				
2009	12.7%	14.6%	7.4%	10.7%
2015	18.1%	19.0%	9.9%	14.6%
Change: 2009–2015	42%	30%	33%	36%
Rented % of land				
2009	8.1%	8.8%	4.5%	6.5%
2015	13.1%	12.8%	6.3%	9.8%
Change: 2009–2015	63%	46%	40%	50%
% HHs with certified land				
2009	6.6%	8.1%	10.2%	9.0%
2015	9.3%	15.9%	13.4%	13.8%
Change: 2009–2015	41%	97%	31%	54%
Rental rates (real 2015 TSh)				
2009	75	75	90	80
2015	126	102	110	108
Change: 2009–2015	68%	36%	22%	36%
Estimated land values (real 2015 TSh)				
2009	1217	1343	1426	1369
2015	1688	2486	2569	2421
Change: 2009–2015	39%	85%	80%	77%

Notes: Calculations are for the rural households in 2008–2009 and 2014–2015 survey rounds of the Tanzanian Living Standards Measurement Study – Integrated Surveys on Agriculture (LSMS–ISA) (www.worldbank.org/en/programs/lsms/initiatives/lsms-ISA, accessed 26 November 2020) data. The sample was refreshed in 2014–2015.

from Mucwini, Uganda, has a slightly different perspective:

> I don't think it matters if you are a man or woman, as long as you have the money to pay the landowner in time. But it is better if a man goes ahead to make the arrangement as men are more respected than the women when it comes to land matters.

Perhaps to circumvent this apparent lack of respect, but also to reduce the need for women to travel away from home, a 26-year-old woman in Awach, Uganda, spoke of how land access is managed among her siblings:

> We use customary land, which is in Latwong, but the land is not enough for all of us. However, my three brothers combined resources and bought some three acres in Latwong, and any female family member is free to use it as long as it is not in use by another person. Meanwhile, the boys rent land from elsewhere while we, the girls, use the [land] at home.

In addition to gender, one's origins may come into play in land rental transactions. For example, according to a 27-year-old woman in Oba Oke, Nigeria, and contrary to what might be assumed, it is the indigenes who are discriminated against, although any positive discrimination seems to come at a cost to the immigrants:

> Most immigrants have access to large size of lands because they have money to pay landowners. Most people love to rent farmland to the migrants, so they have more opportunities than us indigenes. Also at harvest, compared to the indigenes, the immigrants give more crops as a 'bonus' to landowners.

Table 4.4. Means of access to farmland by interviewees.

Country	Site	Status[a]	Cases	% Family, customary or borrowed	% Rented	% Owned	% Sharecropped
Uganda	Awach	Rel. hot	27	88	11	0	0
	Luweero	Rel. hot	36	58	25	17	0
	Mucwini	Rel. cold	42	69	26	5	0
	Butuntumula	Rel. cold	23	52	30	17	0
Ethiopia	Wodo Genet	Rel. hot	14	86	7	0	7
	Bora	Rel. hot	20	15	35	35	15
	Jabi Tehnan	Rel. cold	29	24	38	17	21
	Kuyu	Rel. cold	13	46	23	0	31
Nigeria	Oba Oke	Rel. hot	16	63	38	0	0
	Igbariam Eziafor	Rel. hot	–	–	–	–	–
	Idi Amu	Rel. cold	10	80	10	10	0
	Umumbo	Rel. cold	13	62	0	38	0
Côte d'Ivoire	Soubré	Rel. hot	7	86	0	14	0
	Odienné	Rel. hot	11	64	4	36	0
	Daloa	Rel. cold	12	58	0	42	0
Uganda		Rel. hot	63	73	18	9	0
		Rel. cold	65	61	28	11	0
Ethiopia		Rel. hot	34	51	21	18	11
		Rel. cold	42	35	31	9	26
Nigeria		Rel. hot	16	63	38	0	0
		Rel. cold	23	71	5	24	0
Côte d'Ivoire		Rel. hot	18	75	2	25	0
		Rel. cold	12	58	0	42	0

[a]'Relatively hot' and 'Relatively cold' refer to the level of agricultural commercialization (see Chapter 1, this volume).

The importance of social relations within these markets was highlighted in two interviews. A 25-year-old man in Bora, Ethiopia, noted the importance of family relations:

> Before my marriage my family gave me only one *qert* [0.25 ha] but later they then gave me an additional *qert*. I also rent in two *qert* of land from my mother-in-law at very low price.

In another example, a 22-year-old immigrant woman from Oba Oke, Nigeria, told us how her brother's good connections and high standing helped her in finding land to rent: 'People have a lot of confidence in him so when he introduces a new person there is no problem.'

A number of interviewees drew attention to the limitations and insecurities associated with land rental. For example, limitations on the crops that can be grown, and underhand behaviour on the part of the landowners are highlighted by a 35-year-old woman from Butuntumula, Uganda:

> When I got married, I found no land to cultivate at my husband's place. I was not used to hiring land, but we needed food for the family. We had to hire land where we could grow maize, beans, cassava and sweet potatoes. [...] I wanted to plant bananas but could not. Where I rent, you are allowed to only cultivate annual crops. Bananas go deeper and the plantation lives for over 5 years. Once you grow bananas, you are perceived to be interested in taking the landlord's land. My fields are now a quarter of an acre. The landlord kept giving different portions to different interested tenants so that he makes a lot of money on the same land.
> I used to have a big area rented for 30,000 shillings per season but the size has kept reducing every season. The landlord reduces the size but does not reduce on the rent. Many people lack a place to cultivate so the demand for land is high now.

A 25-year-old woman from Luweero, Uganda, who cultivates cassava, maize and beans on two acres of rented land, told of a very different experience with her landlord, but faces similar restrictions on the crops that she can grow:

> The landlord is a kind man and very patient with us. There are not so many restrictions on his land, except that we cannot grow crops that take long in the garden, those crops that take years to mature.

A 22-year-old male immigrant from Benin, who is farming in Oba Oke, Nigeria, spoke of how through renting one can become entangled in local land disputes:

> [...] a person has three or maybe five children, grows old and then dies. So, the land that '*le vieux*' ['the old man'] possessed now goes to those children, and some of them might not be living in or even visiting the village. It is the children who are in the village that 'attack' the land and who might rent out some of it to you. It can now happen that the children not living in the village return, and insist that the land is for all the children, so you can become involved in a dispute. Then you will have to beg them to allow you to finish with the crops you have started. They might accept, but if not, you might have to give the person something to calm him. And if this does not help, they might send you off the land.

Finally, the experience of a 26-year-old married woman from Jabi Tehnan, Ethiopia, shows that even with the protection provided by a written contract, functional institutions for dispute resolution are essential:

> We contracted farmland for 8400 birr for 2 years. We used it for 1 year, half the time of the contract. The landholder then said he wanted to increase the rent to 10,000 birr for the second year, and we should sign a new 2-year contract. There are people who work as brokers and they encouraged him to increase the rent. We became angry and asked him why he was asking us for another contract with more money before the original contract had finished. We said that he should remember how we are helping him, but he insisted. We took the money on credit from the Amhara Saving and Credit Institute, and we asked him to pay the interest if he is going to break the contract. He should also have paid the *giyyid* [penalty in cash if she/he breaks the contract] but he refused even when he was asked by the elders. The good thing is that we made a written agreement at the beginning. He started intimidating us. [...] He went to my husband's parents and told them that he would kill their son, my husband, that very night. My husband is a religious person and keeps control of himself; I am rather quick to take offence and tried to confront the landlord. [...] He then took the case to a court and I was called to the court to reply to the charge. [...] We took the eyewitnesses [to the original contract] to testify, but by the time we reached the court, he had already been inside with people as eyewitnesses who were not even present during the contract process. We complained and asked the court to consider the contractual agreement and the eyewitnesses indicated on the paper. The landlord started begging the eyewitnesses not to give testimony against him. He also continued intimidating me and my husband. I complained to the court about the harassment. Finally, he returned our 4400 birr [...].

In synthesizing the available evidence on young people and land markets in Africa, while it is clear that markets are increasingly important, there is a wide range of engagement strategies and outcomes. There is significant spatial heterogeneity in these patterns, and markets are not equally important everywhere. Where they are important, they are particularly so for young people. Paradoxically, given the relatively greater importance, there seem also to be higher costs for them to access such markets. In some situations, for example where young people may seek to grow vegetables for the market, these higher costs might be related to the quality of the land being rented (e.g. relatively fertile land in lowland areas where water is available for irrigation). Qualitative accounts of evolving land institutions indicate how difficult it is to cleanly describe the nature of youth participation or assert the impacts of land commodification on young people's livelihoods. The lack of well-defined sampling frames and small sample sizes mean that it is difficult to know how prevalent the different stories and situations are in the broader national, regional or subcontinental contexts. More comprehensive mixed methods studies of the costs and benefits of land markets would be extremely valuable, as would more comprehensive assessments of other dimensions of land commodification and its implications for the livelihoods of rural women and men.

Changes in farm structure are associated with evolving rural economic opportunities

Closely linked to the discussion of commodification and rental markets is the growth in medium-scale farms that, in recent years, has received considerable research attention (Jayne *et al.*, 2016, 2019). How this trend will impact young people will depend on the contribution these larger farms make to the rate of agricultural productivity growth and the strength of their forward and backward linkages with the rest of the economy (see Chapter 3, this volume). To the extent that agricultural productivity growth and new linkages support the creation of off-farm jobs, the increased difficulty of acquiring land for farming, although problematic, may not depress rural livelihoods as would otherwise be expected.

Emerging evidence points to potential positive spillover effects from medium-scale farms on neighbouring small farms (Deininger and Xia, 2016; Lay *et al.*, 2018; Chamberlin and Jayne, 2020), although whether or not such changes in farm structure imply greater employment options for rural youth is not yet clear. Chamberlin and Jayne (2020) enumerate several potential pathways that could generate positive spillovers. First, larger farms which are more highly capitalized and linked with input markets, may help to attract input suppliers to an area and improve upstream linkages for neighbouring smallholders. Second, larger farms are more likely to have a commercial orientation (i.e. sell a larger share of their production) and to market in larger volumes. These characteristics may attract more competition in downstream marketing channels (e.g. buyers, transporters), including larger buyers who may also offer neighbouring smallholders more favourable and well-defined prices. There is some evidence that this is occurring in Zambia (Sitko *et al.*, 2018). Third, if larger farms are owned and managed locally, then there may be relatively high expenditures on local goods and services, i.e. the classical multiplier effects of Johnston and Mellor (1961). Fourth, there may be direct economic engagements between larger and smaller farms, such as larger farms hiring labour from small farm households, or providing tractor services or equipment rentals to neighbouring smallholders. A fifth possible mechanism is via the local stock of political capital: if larger farms are politically important to incumbent administrations, then resource allocation (e.g. for infrastructure and public services) may favour areas in which they are concentrated. Further inquiry into the employment effects arising from the increasingly capitalized and larger-scale farming systems is now warranted.

Land Access is a Key Conditioner of Other Processes

In addition to directly facilitating agriculture, access to land shapes other options available to, and decisions made by, young people. Constrained access to farmland (whether perceived or actual) is associated with the greater likelihood of rural out-migration by young people (Bezu and Holden, 2014; Muyanga and Jayne, 2014; Kosec *et al.*, 2016; Ghebru *et al.*, 2018). Evidence from Kenya also indicates that the amount of land controlled by youth is more important than family land access in influencing their migration decisions (Muyanga *et al.*, 2020).

Similarly, constrained access to land is associated with movement into non farm activities. Table 4.5 presents the estimated results from a fractional Probit model examining the effect of land access on the proportion of total work time devoted to agriculture. The results show that access to land, as proxied by a household's landholdings, significantly influence how young people allocate their labour between farming and non-farm employment activities, even after controlling for individual-, household- and community-observable characteristics. Specifically, for both youth and young adults, increasing household landholdings are associated with greater engagement in farming. This finding is consistent with other studies showing that in at least some locations, the availability of land through inheritance, as well as the productivity of that land, significantly influence the intention of youth to remain in agriculture (Kosec *et al.*, 2018; Mdoe *et al.*, 2020).

The results also provide some insights into how land concentration may be influencing youth engagement in agriculture. Generally, communities with increased concentration of medium-scale farms (5–10 ha) were associated with an increased level of engagement in farming activities among young people. This could signal the role played by medium-scale farms in providing employment to youth, whether directly in agricultural wage work, or indirectly through positive spillover effects on neighbouring small farms such as improved access to market and farm

Table 4.5. Effect of household landholdings on young people's engagement in farming. From Yeboah *et al.*, 2019, based on estimates from a fractional Probit CRE model using data from Tanzania National Panel Survey (2008–2009, 2010–2011, 2012–2013).

	Youth (age 15–24)	Young adults (age 25–35)	Combined (age 15–35)
Household member level			
Age of member (years)	0.2053	0.00	0.0835
	(0.02)***	(0.02)	(0.007)***
Gender of member (1=male)	−0.0152	−0.20	−0.1646
	(−0.11)	(0.10)**	(0.07)**
Member's education attainment (base: no education)			
Primary education completed	−0.8966	−0.28	−0.5825
	(0.15)***	(0.12)**	(0.10)***
Secondary education completed	−2.7340	−1.24	−2.0832
	(0.19)***	(0.19)***	(0.13)***
Post–secondary education completed	−39.8565	−14.49	−74.5801
	(0.31)***	(0.36)***	(0.23)***
Household level			
Age of household head	−0.0136	0.00	−0.0095
	(0.004)***	(0.004)	(0.003)***
Sex of household head (1=male)	0.2293	−0.24	−0.0241
	(0.19)	(0.20)	(0.13)
Marital status (base: monogamous)			
Polygamous	−0.2445	−0.07	−0.1556
	(0.16)	(0.15)	(0.10)
Single	0.2583	−0.43	−0.0540
	(−0.20)	(0.22)**	(0.14)
Other	0.3806	−0.05	0.1165
	(0.18)**	(0.14)	(0.11)
Landholding (ha)	0.2099	0.13	0.1594
	(0.05)***	(0.03)***	(0.03)***
Number of livestock '000	−0.005	0.00	−0.0043
	(0.002)**	(0.003)	(0.001)***
Own plough (1=yes)	0.1266	0.17	0.2013
	(0.17)	(0.20)	(0.12)
Community level			
% of households with 5–10 ha of land	0.0273	0.03	0.0240
	(0.008)***	(0.008)***	(0.006)***
% of households with over 10 ha of land	0.0000	0.00	0.0021
	(0.004)	(0.004)	(0.003)
Residuals from first stage regression	−0.1797	0.04	−0.0204
	(0.07)**	(0.04)	(0.04)
Constant	−3.8413	0.72	−1.4963
	(0.51)***	(0.58)	(0.27)

Notes: Results from first stage regression controlling for factors influencing household landholding sizes are not reported here. Standard errors and significance levels in parentheses; significance levels denoted as follows: * = p<0.10, ** = p<0.05, *** = p<0.01.

inputs, which increases profitability of farming and creates new incentives.

Lack of available land has been implicated in delayed household formation and associated with so-called 'waithood' (Honwana, 2014). As discussed in Chapter 8 (this volume), however, it would be a mistake to draw too narrow a connection between limited access to land, low labour market participation and waithood. Young people work to sustain themselves and their families, and to attain the markers of social adulthood through avenues other than autonomous farming, wage work and/or self-employment.

Given the important theoretical linkage between long-term tenure security and incentives for productive investments which are not fully realizable within a season, the increased shift to production on rented or borrowed land may be associated with greater nutrient mining or short-term investments, over longer-term maintenance of soil fertility. While some research suggests that this may be the case (Ricker-Gilbert *et al.*, 2020), evidence is still limited. Still the orientation of agricultural investments by young farmers may be especially sensitive to incentives of production under short-term usufruct rights (e.g. Okali and Sumberg, 2012).

Given the cash needs for upfront rental contracts (the most prevalent rental arrangement in SSA), the role of non-farm occupations in enabling farming activities must be considered. The cash needs of commercially oriented farming may further underscore the importance of capital resources. This diverges from the age-old pathway of 'pulling oneself up by one's bootstraps' (with help from family and friends), without recourse to resources outside of one's own labour and capital. Little empirical work to date has explored this relationship, but an interview with a 25-year-old man from Jabi Tehnan, Ethiopia, provided an intriguing example:

> I also started to rent in land from the land-rich people to do farming. I used to rent in one *gazim* [0.25 ha] of land for 1 year paying 3000–4000 birr to cultivate green pepper. The main activity that is important for my livelihood is obviously green pepper production. Regardless of the high cost of production, including buying chemical fertilizers, and limited profit, my daily subsistence depends on the income from producing and selling pepper. I produce it on the land that I rent in from other people. […] It [money for renting farmland] is obtained from the teahouse. I work in the teahouse and generate the money to rent in land for farming.

Opening New Empirical Windows on Remaining Knowledge Gaps

The foregoing discussion indicates that nationally representative household survey data are invaluable for monitoring changes in Africa's rural land access and use – in terms of the distribution of farm sizes, evolving land institutions and the patterns of farming. However, the empirical windows they offer have important limitations for developing a youth-specific lens on land dynamics.

To begin with, standard survey data offer constrained windows into processes of household formation. It is difficult to determine where any given surveyed household falls within this process. Even survey questions such as, 'when did the household head first come to this area?', do not indicate whether the currently observed household formed at that time, prior to the head's arrival, or at some point after arrival. Moreover, traditional household survey instruments are subject to the tyranny of the conception of the household as an organizational unit, and the household head as the locus of farm and household management. As a result it is entirely possible that the activities of young dependents (farming, labour allocations, business activities, etc.) do not show up at all in survey responses. Young people may occupy liminal or interstitial places that may be opaque to surveys focused on the household strongly framed relative to the head. For example, given the low reported rates of plot decision making and control by household members other than the head and spouse (see Chapter 6, this volume), it seems plausible that we may not fully observe young people's land management activities. The divergence between the nominal patterns of intrahousehold control as recorded in survey data, and the self-expressed narrative accounts of youth farming engagement (e.g. Andersson Djurfeldt *et al.*, 2019, as well as the qualitative findings reported above) suggest that closer attention to this is warranted. Such patterns may stem from systematic survey omissions (i.e. failing to record data on plots which are not controlled by the head or spouse) and/or divergent subjective interpretations of control. Either way, digging deeper into the ways in which dependents farm (and whether or not such activities show up in household survey plot rosters) is clearly warranted.

Finally, given the dynamism around both land and youth in at least some parts of rural SSA, there may be a rationale for surveys to oversample relatively dynamic contexts. Examples of such areas may include: commercialization hot spots, farm blocks (which hire in labour), areas undergoing rapid changes in urbanization or accessibility, or areas where farming is changing significantly (e.g. via expansion of medium- and larger-scale farms) and value chains are rapidly modernizing.

Conclusions

In this chapter, we have attempted to articulate why an understanding of rural youth livelihoods in SSA requires an appreciation of what is happening with land. While land-related research questions are not necessarily always youth questions, youth-related questions almost invariably invoke consideration of land, either directly or indirectly. In our discussion, we have endeavoured to do three things. First, we summarize recent empirical evidence from primary and secondary sources to sketch out what we know about youth and land in rural Africa. While many aspects of the stylized story are clearly borne out in their broad contours from existing empirical data – e.g. the increasing importance of land rental, shrinking farm sizes – other aspects require more nuance, such as the supposed disinterest in rural farming (and non-farming) futures on the part of young people. Second, we review how land questions spill over into other youth options, decisions and outcomes, through a review of recent analyses. Building on these discussions, a third contribution of this chapter is to articulate remaining knowledge gaps and to discuss the ways that new modes of data collection could help address these gaps.

Africa is undergoing rapid transformation with respect to many of the land-related questions and characteristics outlined above (Jayne *et al.*, 2018). Rural population growth is expected to continue for several more decades, even as urbanization increases (United Nations, 2019b). This means that new generations of rural young people coming into the rural workforce will feed into greater absolute numbers of rural households and individuals. Land constraints will inevitably be more acute for these successive generations. Livelihood prospects for these young people will be dependent upon, on the one hand, how they are able to negotiate the increasing commodification of land and evolving land institutions (markets, tenure formalization) and, on the other, the kinds of opportunities that are opened up off-farm in increasingly capitalized and market-oriented farming systems and rural economies.

As we have discussed, an improved and more nuanced understanding of these issues as 'youth issues' will require new approaches to data collection and analysis. In particular, ways need to be found to better monitor the interstitial spaces occupied by youth that may not show up in survey sampling frames (or in conventional modes of household-based data collection). We envisage much more scope for exploratory qualitative work to guide better and novel sampling frames, better questions and modes of survey data collection, as well as greater scope for intentional mixed methods.

Notes

[i] In Tanzania, this includes two main types of certificates – a Granted Right of Occupancy (GRO) issued by the government, or a Customary Right of Occupancy (CRO) issued by a village council – as well as derivative rights of a GRO, such as leases and other private contractual arrangements which may be legally formalized.

[ii] Data on two types of land values are available from LSMS–ISA surveys in Tanzania: data on actual rental rates paid by tenants (for plots which were rented-in); and estimated sales values for land, elicited as responses to the question, 'What would the value of this plot be if it were sold today?'

[iii] The interviews relied on people's own understanding of ownership. We assume that this can mean either the land was purchased (with or without formal title) or held through long-term customary usufruct rights which may or may not extend to its sale.

[iv] See Yeboah *et al.* (2020) for an example of youth sharecropping in Ghana involving the establishment of a cashew plantation.

Acknowledgements

We are grateful to Kibrom Abay, Woinishet Asnake and Hailemariam Ayalew for contributing to the quantitative analyses for Burkina Faso, Ethiopia, Uganda, Niger and Nigeria.

References

Abay, K., Berhane, G. & Chamberlin, J. 2021. Are Land Rental Markets Responding to Rising Population Pressures and Land Scarcity in sub-Saharan Africa? Land Use Policy, 101 (doi.org/10.1016/j.landusepol.2020.105139)

Amanor, K. S. 2012. Global resource grabs, agribusiness concentration and the smallholder: two West African case studies. *The Journal of Peasant Studies,* 39, 731–749.

Amanor, K. S. 2018. Markets, politics and land administrative reform in Africa: what can African studies contribute? *In:* Kirikoshi, H., Matsunami, Y., Takeuchi, S. & Midorikawa, N. (eds.) *Frontiers of African Studies – TUFS 'Kick-off' Symposium, 2018*, Tokyo, African Studies Center – Tokyo University of Foreign Studies (ASC – TUFS), 39–52.

Andersson Djurfeldt, A., Kalindi, A., Lindsjö, K. & Wamulume, M. 2019. Yearning to farm – Youth, agricultural intensification and land in Mkushi, Zambia. *Journal of Rural Studies,* 71, 85–93.

Anseeuw, W., Jayne, T., Kachule, R. & Kotsopoulos, J. 2016. The quiet rise of medium-scale farms in Malawi. *Land,* 5, 19.

Asaaga, F. A. & Hirons, M. A. 2019. Windows of opportunity or windows of exclusion? Changing dynamics of tenurial relations in rural Ghana. *Land Use Policy,* 87 (https://doi.org/10.1016/j.landusepol.2019.104042).

Bezu, S. & Holden, S. 2014. Are rural youth in Ethiopia abandoning agriculture? *World Development,* 64, 259–272.

Chamberlin, J. & Jayne, T. S. 2020. Does farm structure affect rural household incomes? Evidence from Tanzania. *Food Policy,* 90, 101805.

Chamberlin, J. & Ricker-Gilbert, J. 2016. Participation in rural land rental markets in sub-Saharan Africa: who benefits and by how much? Evidence from Malawi and Zambia. *American Journal of Agricultural Economics,* 98, 1507–1528.

Chinsinga, B. & Chasukwa, M. 2012. Youth, agriculture and land grabs in Malawi. *IDS Bulletin,* 43, 67–77.

Chitonge, H., Mfune, O., Kafwamba, D. & Kajoba, G. 2017a. Hybrid land markets: monetarised customary land transactions in Zambia. *Canadian Journal of African Studies / Revue Canadienne des Etudes Africaines,* 51, 123–143.

Chitonge, H., Mfune, O., Umar, B. B., Kajoba, G. M., Banda, D. & Ntsebeza, L. 2017b. Silent privatisation of customary land in Zambia: opportunities for a few, challenges for many. *Social Dynamics,* 43, 82–102.

Cousins, B. 2007. More than socially embedded: The distinctive character of 'Communal Tenure' regimes in South Africa and its implications for land policy. *Journal of Agrarian Change,* 7, 281–315.

CSO 2012. *The 2012 Rural Agricultural Livelihoods Survey (RALS) Interviewer's Instruction Manual. Central Statistical Office (Agriculture and Environment Division) in conjunction with the Ministry of Agriculture and Livestock (Policy and Planning Department), commissioned by Michigan State University (MSU) and Indaba Agricultural Policy Research Institute (IAPRI),* Lusaka, Zambia, Central Statistical Office (CSO).

Deininger, K., Savastano, S. & Xia, F. 2017. Smallholders' land access in sub-Saharan Africa: A new landscape? *Food Policy,* 67, 78–92.

Deininger, K. & Xia, F. 2016. Quantifying spillover effects from large land-based investment: The case of Mozambique. *World Development,* 87, 227–241.

Ghebru, H. 2019. *Perceived Tenure (In)security in the Era of Rural Transformation: Gender-disaggregated Analysis from Mozambique. Feed the Future Innovation Lab for Food Security Policy Research Paper 125*, East Lansing, MI, Michigan State University.

Ghebru, H., Amare, M., Mavrotas, G. & Ogunniyi, A. 2018. *Role of Land Access in Youth Migration and Youth Employment Decisions: Empirical Evidence from Rural Nigeria. Strategy Support Program Working Paper 58,* Washington, DC, IFPRI.

Hall, R., Scoones, I. & Tsikata, D. 2017. Plantations, outgrowers and commercial farming in Africa: agricultural commercialisation and implications for agrarian change. *The Journal of Peasant Studies,* 44, 515–537.

Holden, S. T. & Bezu, S. 2016. Preferences for land sales legalization and land values in Ethiopia. *Land Use Policy,* 52, 410–421.

Holden, S. T. & Tilahun, M. 2018. *Land Rental as a Complementary Income Source for Land-poor Youth. CLTS Working Papers 6/18,* Aas, Norway, Norwegian University of Life Sciences, Centre for Land Tenure Studies.

Holden, S., Otsuka, K. & Place, F. (eds.) 2008. *The Emergence of Land Markets in Africa: Impacts on Poverty, Equity, and Efficiency,* London, Earthscan.

Holden, S. T., Otsuka, K. & Deininger, K. (eds.) 2013. *Land Tenure Reforms in Asia and Africa: Impacts on Poverty and Natural Resource Management,* London and New York, Palgrave Macmillan.

Honwana, A. 2014. "Waithood": youth transitions and social change. *In:* Foeken, D., Dietz, T., Dd Haan, L. & Johnson, L. (eds.) *Development and Equity: An Interdisciplinary Exploration by Ten Scholars from Africa, Asia and Latin America,* Leiden, Netherlands, Brill.

Jayne, T. S., Chamberlin, J. & Benfica, R. 2018. Africa's unfolding economic transformation. *Journal of Development Studies,* 54, 777–787.

Jayne, T. S., Chamberlin, J. & Headey, D. D. 2014. Land pressures, the evolution of farming systems, and development strategies in Africa: a synthesis. *Food Policy,* 48, 1–17.

Jayne, T. S., Chamberlin, J., Traub, L., Sitko, N., Muyanga, M., Yeboah, F. K., Anseeuw, W., Chapoto, A., Wineman, A. & Nkonde, C. 2016. Africa's changing farm size distribution patterns: the rise of medium-scale farms. *Agricultural Economics,* 47, 197–214.

Jayne, T. S., Muyanga, M., Wineman, A., Ghebru, H., Stevens, C., Stickler, M., Chapoto, A., Anseeuw, W., Van Der Westhuizen, D. & Nyange, D. 2019. Are medium-scale farms driving agricultural transformation in sub-Saharan Africa? *Agricultural Economics,* 50, 75–95.

Jayne, T. S., Yamano, T., Weber, M. T., Tschirley, D., Benfica, R., Chapoto, A. & Zulu, B. 2003. Small-holder income and land distribution in Africa: implications for poverty reduction strategies. *Food Policy,* 28, 253–275.

Jin, S. & Jayne, T. S. 2013. Land rental markets in Kenya: implications for efficiency, equity, household income, and poverty. *Land Economics,* 89, 246–271.

Johnston, B. F. & Mellor, J. W. 1961. The role of agriculture in economic development. *The American Economic Review,* 51, 566–593.

Kidido, J. K., Bugri, J. T. & Kasanga, R. K. 2017. Dynamics of youth access to agricultural land under the customary tenure regime in the Techiman traditional area of Ghana. *Land Use Policy,* 60, 254–266.

Knapman, C., Silici, L., Cotula, L. & Mayers, J. 2017. *Africa's Farmland in Changing Hands,* London, International Institute for Environment and Development (IIED).

Kosec, K., Ghebru, H., Holtemeyer, B., Mueller, V. & Schmidt, E. 2016. *The Effect of Land Inheritance on Youth Employment and Migration Decisions: Evidence from Rural Ethiopia,* Washington, DC, International Food Policy Institute (IFPRI).

Kosec, K., Ghebru, H., Holtemeyer, B., Mueller, V. & Schmidt, E. 2018. The effect of land access on youth employment and migration decisions: evidence from rural Ethiopia. *American Journal of Agricultural Economics,* 100, 931–954.

Kumeh, E. M. & Omulo, G. 2019. Youth's access to agricultural land in sub-Saharan Africa: A missing link in the global land grabbing discourse. *Land Use Policy,* 89 (https://doi.org/10.1016/j.landusepol.2019.104210).

Lawry, S., Samli, C., Hall, R., Leopold, A., Hornby, D. & Mtero, F. 2014. *The Impact of Land Property Rights Interventions on Investment and Agricultural Productivity in Developing Countries: A Systematic Review,* Oslo, The Campbell Collaboration.

Lay, J., Nolte, K. & Sipangule, K. 2018. *Large-Scale Farms and Smallholders: Evidence from Zambia. Working Paper 2098,* Kiel, Germany, Kiel Institute for the World Economy.

Mdoe, N. S. Y., Muyanga, M. C., Jayne, T. S. & Minde, I. J. 2020. Factors influencing young Tanzanians to stay in farming or migrate out of rural areas (Unpublished paper).

Muraoka, R., Jin, S. & Jayne, T. S. 2018. Land access, land rental and food security: Evidence from Kenya. *Land Use Policy,* 70, 611–622.

Muyanga, M., Jayne, T. S. 2014. Effects of rising rural population density on smallholder agriculture in Kenya. *Food Policy*, 48, 98–113.

Muyanga, M., Otieno, D. & Jayne, T. S. 2020. Pulled or pushed out? Causes and consequences of youth migration from densely populated areas of rural Kenya. Unpublished paper.

Okali, C. & Sumberg, J. 2012. Quick money and power: Tomatoes and livelihood building in rural Brong Ahafo, Ghana. *IDS Bulletin,* 43, 44–57.

Ricker-Gilbert, J. & Chamberlin, J. 2018. Transaction costs, land rental markets, and their impact on youth access to agriculture in Tanzania. *Land Economics,* 94, 541–555.

Ricker-Gilbert, J., Chamberlin, J., Kanyamuka, J., Jumbe, C. B. L., Lunduka, R. & Kaiyatsa, S. 2019. How do informal farmland rental markets affect smallholders' well-being? Evidence from a matched tenant-landlord survey in Malawi. *Agricultural Economics,* 50, 595–613.

Ricker-Gilbert, J, Chamberlin, J, and Kanyamuka, J. (2018) Wealth Disparities and Soil Fertility on Rented versus Owned Plots: Evidence from a Matched Tenant/Landlord Sample in Malawi, 2018 Annual

Meeting, August 5–7, Washington, D.C., Agricultural and Applied Economics Association. (doi.org/10.22004/ag.econ.274237).

Sitko, N. & Chamberlin, J. 2015. The anatomy of medium-scale farm growth in Zambia: what are the implications for the future of smallholder agriculture? *Land,* 4, 869–887.

Sitko, N. J. & Chamberlin, J. 2016. The geography of Zambia's customary land: Assessing the prospects for smallholder development. *Land Use Policy,* 55, 49–60.

Sitko, N. J., Burke, W. J. & Jayne, T. S. 2018. The quiet rise of large-scale trading firms in East and Southern Africa. *The Journal of Development Studies,* 54, 895–914.

The World Bank 2020. World development indicators. The World Bank Data Bank (https://data.worldbank.org/region/sub-saharan-africa) accessed 26 January 2021.

Tione, S. E. & Holden, S. T. 2019. *Transaction Costs and Land Rental Market Participation in Malawi. CLTS Working Papers 9/19,* Aas, Norway, Norwegian University of Life Sciences, Centre for Land Tenure Studies.

Tione, S. E. & Holden, S. T. 2020. Urban proximity, demand for land and land shadow prices in Malawi. *Land Use Policy,* 94, 104509.

United Nations 2019a. *2018 Demographic Yearbook, Sixty-ninth Issue,* New York, United Nations.

United Nations 2019b. *World Urbanization Prospects 2018. (ST/ESA/SER.A/421),* New York, United Nations, Department of Economic and Social Affairs, Population Division.

White, B. 2012. Agriculture and the generation problem: rural youth, employment and the future of farming. *IDS Bulletin,* 43, 9–19.

Wineman, A. & Jayne, T. S. 2018. Land prices heading skyward? An analysis of farmland values across Tanzania. *Applied Economic Perspectives and Policy,* 40, 187–214.

Yaro, J. A., Teye, J. K. & Torvikey, G. D. 2017. Agricultural commercialisation models, agrarian dynamics and local development in Ghana. *The Journal of Peasant Studies,* 44, 538–554.

Yeboah, F. K. & Jayne, T. S. 2018. Africa's evolving employment trends. *Journal of Development Studies,* 54, 803–832.

Yeboah, F. K., Jayne, T., Muyanga, M. & Chamberlin, J. 2019. *Youth Access to Land, Migration and Employment Opportunities: Evidence From sub-Saharan Africa. IFAD Research Series 53*, Rome, International Fund for Agricultural Development (IFAD).

Yeboah, T., Chigumira, E., John, I., Anyidoho, N. A., Manyong, V., Flynn, J. & Sumberg, J. 2020. Hard work and hazard: Young people and agricultural commercialisation in Africa. *Journal of Rural Studies,* 76, 142–151.

5 Mobility and the Rural Landscape of Opportunity

Dorte Thorsen[1] and Thomas Yeboah[2]
[1]*Institute of Development Studies, Brighton, UK;* [2]*Bureau of Integrated Rural Development (BIRD), Kwame Nkrumah University of Science and Technology (KNUST), Kumasi, Ghana*

Introduction

Migration features prominently in policy discourse relating to rural areas in sub-Saharan Africa (SSA) and has done for many decades. The framing of rural out-migration as a policy problem usually highlights the threat to the agricultural sector and national food security from the loss of young labour; the threat to urban areas from additional pressure on jobs and services; and/or the threat to young people themselves as they are pushed into risky, anti-social, criminal or radicalized behaviour. All three framings suggest that for the sake of society, the economy and rural young people, the policy objective must be to limit rural out-migration.

Against this background, over the last two decades there has been a turn within the social sciences toward what is called a 'mobilities paradigm' (Benwell, 2009). The rationale is that because populations in Africa have historically been mobile – relocating nearby or internationally, for shorter or longer periods of time, individually or in groups – the way movement through space is talked about varies, rendering the term 'migration', and the theoretical underpinnings of much migration research, highly problematic. Centring the analysis on mobilities overcomes sticky issues of whether or not relocation to a nearby village, an adjoining district, or the other side of the country constitutes migration. Moreover, it allows for a more flexible understanding of the nexus between involuntary and voluntary relocation.

The mobilities paradigm has ushered in new studies on the livelihoods and geographies of young people, taking account not only of migration, but also shorter distance and shorter duration movements (Hannam *et al.*, 2006). The shift from migration to mobilities has also brought increased critical attention to the interplay between mobilities and youth transitions, with Crivello (2011) arguing for the need to situate young people's spatial mobilities in relation to transitions around school, work, personal relationships, social adulthood and so on. Growing up, rural young people must negotiate these transitions, and for many, their ability to become mobile helps facilitate this across space and time (Thorsen, 2006; Porter *et al.*, 2010; Punch, 2015). Such negotiations start in childhood. The interplay between mobility and youth transitions is shaped by birth order, age, education, gender, local circumstances, parental status and young people's own marital status, along with broader cultural and political economy contexts. Indeed, while mobility may present opportunities for some, for many others it is a necessity (in their eyes, or the eyes, for example, of parents) (Porter *et al.*, 2008; Hashim and Thorsen, 2011; Punch, 2014); and for still others, it may be of no interest at all.

 Youth and the Rural Economy in Africa: Hard Work and Hazard (ed. J. Sumberg)
DOI: 10.1079/9781789245011.0005

Young people's involvement in different forms of mobility is a response to changing sociocultural and economic circumstances over space and time (Cole, 2010; Veale and Donà, 2014). Mobility is 'engrained in the history, daily life and experiences of people' and as with the whole array of forms of human behaviour, it is 'inspired by different motives, aspirations and obligations' (de Bruijn *et al.*, 2001, p.1). It is imperative to understand how mobility has served, and continues to serve, as an important strategy in young people's livelihood building – directly in terms of accessing employment opportunities or land, and indirectly through access to better education or healthcare, which in turn may help shape future prospects for employment and social mobility. This is equally true for livelihoods involving labour migration to large cities or distant places, transitory or nearby relocations, and/or efforts to build livelihoods involving little if any mobility (Carling and Schewel, 2018; Gaibazzi, 2018).

If youth mobility in SSA continues to attract the attention of researchers, work on the mobility patterns of rural youth and how these relate to transitions and livelihood building is still relatively scarce (Porter *et al.*, 2017). Much of the existing body of research is focused on urban contexts, with specific reference to marginalized street or migrant youth, or the wider cohort of youth (see Gough, 2008, on Lusaka; Benwell, 2009, on Cape Town; Langevang and Gough, 2009, on Accra). For example, poor unemployed youth in urban Accra have been reported to move into and occupy certain spaces over which they then exert some degree of control. In the same context, and with luck, 'moving around' may position young people to take advantage of whatever opportunities present themselves – in a market, on a street corner, outside a supermarket in a middle-class neighbourhood, or further afield (Langevang and Gough, 2009).

In this chapter, we look at young women's and men's strategies for mobility in rural economies in Uganda, Ethiopia, Nigeria and Côte d'Ivoire. We draw on livelihood, life history and photo–voice interviews, as well as focus group discussions (FGDs), with young people across 16 sites in four countries (see Chapter 2, this volume). The chapter focuses on spatial mobilities resulting from: involuntary relocations because of conflict in society and/or disruption within the family setting; relocations for education; and relocations for work. The analysis investigates in particular the gendered nature of youth mobilities and immobilities, and their implications for livelihood building. After reviewing work on rural youth mobilities we conceptualize youth mobilities in terms of aspirations and imagined life projects. The analysis then addresses involuntary relocations, relocations for school and relocations for work. The last section concludes the discussion.

Youth Mobilities, Transitions and Life Projects

Mobility, education and work

Mobility, including migration for work, is 'a *process* and more than a single act or event of relocation' (Crivello, 2011, p.396). Over the years, mobilities of rural people in SSA have included relocation to access education; relocation following marriage; large-scale rural movement in search of arable land for farming; forced movement following civil unrest; and relocation to towns, cities and further afield in search of employment (Hashim and Thorsen, 2011). For rural youth, engagement with one or more of these mobilities is structured by social relations, and is also embodied and highly gendered (Hashim and Thorsen, 2011; Porter *et al.*, 2017). Farrugia (2016, p.838) argues that rural youth mobilities should be seen as embedded in the wider structural processes constituted by capitalist modes of accumulation that underlie spatial inequalities. For many rural young people, mobility is central to survival, transitions and life chances.

Education is an important avenue through which young people negotiate their social and economic positions (see Chapter 7, this volume), and for many, education and mobility are intricately intertwined. Rural youth are often disadvantaged in terms of education quality because of more limited access, the non-availability of teachers, inappropriate curricula and having less time to study (Meinert, 2003; Hashim, 2007; Camfield and Tafere, 2009). However, suggesting that educational disadvantage can be understood solely in terms of rural–urban comparisons, frames relocation for education within a simplistic model of push–pull factors, thus ignoring the more complex and intersecting reasons for mobilities in and through education.

Whether they are attending school or not, as subordinate members of households and kinship groups, young people's labour is integrated into the portfolio of activities, both productive (farming, trading, etc.) and reproductive (sweeping, caring for younger siblings, water and fuel collection, cooking, etc.), that contribute to household and individual livelihoods. Apart from working on the family farm, children and youth are often encouraged to develop their own economic spheres. If land is available, they may be given a plot on which to farm, or encouraged into non-farm activities. In rural savannah regions of West Africa, the long history of youth mobility – for schooling, apprenticeship and/or work – suggests it should be seen as a normal part of young people's lives and therefore of youth transitions (Hashim and Thorsen, 2011; Thorsen, 2014). In addition to daily journeys to school, rural youth may move around locally in order to farm; collect firewood, water and livestock feed; to buy or sell farm produce; and to meet and socialize with peers. Actual movement patterns depend on resource availability as well as on normative views on gender and age-appropriate work, but commonly time and space are given to young people with the expectation that they engage in independent economic activity. This expectation is a significant reason why many rural youth move beyond their immediate villages (Hashim and Thorsen, 2011; Yeboah, 2020).

While both girls and boys participate in productive and reproductive activities, in most rural environments, the heaviest of these household-level work demands fall on young women, and may constrain their mobility and therefore their ability to access educational and economic opportunities. In their study in rural Malawi, Porter *et al.* (2011) found that the mobility of young women of all ages is constrained because of widespread sensitivities around their susceptibility to sexual assault and promiscuity, which only increases with puberty. In contrast, rural boys and young men have more freedom to move beyond their 'local environment' because of lower domestic workloads, less concern about sexual assault, an implicit understanding that they are less controllable, and the perception that the mobility of young men has less impact on family welfare. Young men being indulged in this way may also reflect the family's desire not to alienate them so that they ultimately stay in the local area and attend to their family responsibilities. However in rural northern Ghana and other parts of rural West Africa, the socialization of young women towards marriage, including the expectation that they become the sole provider of household items such as cooking pots and that they should generate their own income, drives their early mobility in search of income generation and procurement opportunities (Hashim, 2005; Hashim and Thorsen, 2011; Yeboah, 2020).

Gough and Birch-Thomsen (2016) draw attention to how in northern Ghana different mobilities enable both young women and young men to build their livelihoods by drawing on experience and capital generated from within their villages (e.g. through farming and fishing). They also describe how the ability to become mobile has enabled young people to exploit more distant opportunities, typically in mining, but also in the urban informal sector, and thus generate further financial resources that could then be invested in assets (e.g. purchase of a motorbike, residential plots and housing) and non-farm businesses. Young people's mobility, in an effort to exploit the landscape of opportunity, is thus fundamental to acquisition of skills, networks, knowledge, experience and financial capital.

In her study in the Bisa region of Burkina Faso, Thorsen (2014) highlights the lack of economic possibilities, fluctuations in income from farming, and not wanting to remain in chronic poverty, as significant drivers of adolescent boys seeking alternative employment opportunities elsewhere, whether the decision is taken by themselves or by/with their parents. Their experience of the rural economy is similar to that of young people in many other contexts across SSA (see also Castle and Diarra, 2003; Crivello and van der Gaag, 2016; Yeboah, 2020). Nevertheless, and notwithstanding economic drivers, a key aspect of young people's mobility is social, and linked to important life course transitions. Young women and men seek more autonomy through their mobility and in so doing, negotiate new independence from – as well as dependencies and interdependencies within – the family and broader kin groups. However, new economic resources and reshaped relationships only contribute to successful transitions and additional social status if their mobility respects prevalent norms of 'good behaviour' (Thorsen, 2006, 2014; Thorsen and Jacquemin, 2015).

Mobility focused research draws attention to the spatial distribution of livelihood opportunities, including farm and non-farm work, in young people's transitions. In the context of constrained opportunity, and high levels of risks and uncertainty, rural people have increasingly become 'occupationally flexible and spatially mobile' (Porter *et al.*, 2008, p.6). Apart from farming activities most often carried out on village lands, where significant mobility may still be required, many other income-generating opportunities require mobility beyond the village boundary. Moreover, for young people from relatively densely populated rural locations, where access to land may be a challenge (see Chapter 4, this volume), relocation to towns and urban areas can be critical for livelihood building, successful transitions (Hashim and Thorsen, 2011) and the accumulation of wealth (Kleist and Thorsen, 2017). Thorsen's (2014) work in rural Burkina Faso shows that mobility enables young men to participate in the global culture of consumerism. In addition to undergoing individuation and enhancing their autonomy and identity, such mobility and the activities it enables help forge intergenerational ties between senior relatives, obligations to family relations and interdependencies that link generations.

Farrugia (2016) suggests that youth mobilities give rise to 'translocal subjectivities' – i.e. engagement with, and temporal connections to, multiple spaces, which young people construct through mobilities. This involves the mobilization of material and symbolic resources across both rural and urban spaces. In effect, 'rural young people's lives can no longer be located purely in one place, but are translocal, or constructed through economic, symbolic and affective relationships between the multiple spaces through which they move' (Farrugia, 2016, p.848).

Social networks

The important role of both peer and adult social networks in the mobility trajectories and transitions of young people, is well documented in both rural and urban locations (e.g. Both, 2010; Porter *et al.*, 2011; Thorsen, 2012). Rural young people tend to depend heavily on these networks as they relocate, and in particular to access employment opportunities and accommodation in unfamiliar contexts. Where these networks involve adults, they may reinforce dependency and asymmetrical power relations (Thorsen, 2012), but they also potentially serve to strengthen interdependencies necessary for the long-term survival and security of the lineage.

Thorsen (2012, p.6) suggests that: 'The thick social relations that engender friendships, reciprocities and some level of responsibilities stretch beyond the family and are key to finding better paid and more secure jobs and thus to upward social mobility.' Poorly paid jobs such as shoe shining or some types of hawking can easily be started independently by young people, and as such require little or no adult mediation. Tetteh (2014) found that rural young people who work as domestics in Ghana exercise agency by personally approaching potential employers in search of jobs, but in so doing they may be compelled to negotiate their own terms of employment and remuneration. While such arrangements give the young people some degree of control over their earnings, inexperience and limited bargaining power may result in unfavourable or even exploitative terms and conditions. Nevertheless, many still find innovative strategies to cope with their situations or change their employers (Tetteh, 2014).

Rural Youth Mobilities and Livelihood Building

Involuntary relocations

The line between relocation being involuntary or voluntary is ambiguous, and sometimes results in choosing the lesser of two evils. The distinction is further complicated because it brings into play power relations between adults, and between adults and young people. Their assessments of the options available are not necessarily the same, and what is deemed preferable by persons with more power to make decisions, may preclude objections from those relocating, and thus bring into question whether mobilities are voluntary. In times of conflict and violence, the room for disagreement and negotiation around relocation shrinks, but choices are nevertheless made.

All four countries in focus in this chapter have experienced some level of political conflict but protracted conflicts involving military and armed rebels, flight and forced relocation have affected Uganda and Côte d'Ivoire most severely. Both countries experienced recurrent armed conflicts: Uganda from the early 1990s to around 2010; and Côte d'Ivoire from 2002 until 2007 and then again in 2010, affecting young people's lives from early childhood into their twenties.

In Uganda, villagers moved into camps for internally displaced persons (IDPs) for protection against the Lord's Resistance Army (LRA); neither adults nor youth had much influence on these decisions in an atmosphere of fear and violence. And yet, interviews reveal different strategies for keeping children safe, overcoming the limited options for growing food or earning an income in the camps, and safeguarding access to land and income in the long term. A young man in Awach, northern Uganda, described a common tale of short-distance childhood mobility in war time, weaving together his parents' death, the LRA insurgency, schooling and social networks, to show the importance of both paternal and maternal relatives:

> I grew up in my uncle's home. When my mother lost her life, we went to stay in her home [some 6 km away]. During the insurgency, I stayed in IDP camps but not for long. I started in Awach [40 km away] where I stayed 3 months, then I went and stayed with my aunty for 3 months, and then my uncle's daughter called me to stay with her [some 30 km away], close to the Coo Pe camp in 2001. I stayed until 2004. I was still young and didn't go to school because they feared that the LRA may abduct us. I started primary school near my maternal village in 2005, then came to St. Mauritz Primary School in Gulu, and I started living with my sister. I passed my PLE [Primary Leaving Exam] very well and an NGO [non-governmental organization] paid for my school fees for 1 year; then in Senior 2 it was the headteacher who paid my school fees. While I was studying, my brother from Amuru told me that our land was being entered by people from different clans, [and] that I should go back and live there. [When I moved] I found that Awach Secondary School was far, [so] I decided to leave school to take up farming.

Although some young people moved further away, the mobilities of the vast majority were temporary solutions of a local and flexible nature, which resulted in greater safety. But at the same time, living in camps undermined the livelihoods of most people, as the camps did not generally offer space for farming or other economic activities. Initially, the education of displaced children was disrupted, but over the years, schools emerged in the camps. More boys than girls spoke about having been abducted by the LRA. Upon return they had difficulties going back to school, as a 35-year-old woman in Awach explained:

> In 1998, while I was in Primary 7, the rebels abducted me. I returned after 2 months and tried to go back to school but dropped out because of the stigma of abduction from other children.

None of the young men spoke of going back to school after abduction, unless they had relocated to live with relatives further away. At the time when parents and older siblings began to move back to their villages to farm, NGO support enabled some youth to continue secondary schooling. As was the case for the young man above, relocation driven by a concern for land rights could disrupt education.

In Côte d'Ivoire, the conflict was spurred by an economic downturn and xenophobia, and migrants from northern Côte d'Ivoire and neighbouring countries were under threat of violence and eviction from their farms in the southern part of the country. At the same time, young people in northern areas were primarily affected by the closure of schools and the rupture in circular (national) mobilities. Camps were mostly located in neighbouring countries and none of the young people in our study spoke of camps. Those whose parents or grandparents were migrants spoke about returning to their ancestral lands, as did a 28-year-old man near Daloa, who was born in Benin but had spent his entire childhood in Côte d'Ivoire:

> When the crisis [the war] started in September 2002, I was attending the last year of primary school here in the village. My father had died in 2000 but the rest of my family was living in the hamlet close to our cocoa farm. At first my family joined me in the village compound, then we all went back to Benin. Later, in 2004, my brothers and I came back but my mother stayed in Benin. I came back to finish my schooling.

> This year I went back to Benin to spend 4 months with my mother. Now I'm back [in Côte d'Ivoire] to pursue my projects. What I would like to do, I will do here in Côte d'Ivoire. I'm Beninese but I don't master the conditions there; I master them here, especially here in the west of the country.

A similar pattern of back-and-forth mobility was seen among migrants with roots in Burkina Faso, whose elders had established cocoa farms through sharecropping arrangements, two or three generations back. Large numbers of people fled during the most violent periods in the war, and children were sent back to be safe and to continue schooling (Zongo, 2016). However, many migrant families relocated from their hamlets into large villages or rural towns and in the periods of relative calm, the circulation of migrants continued, both over longer and shorter distances. As with the Beninese youth referred to above, people returned well before the peace accord was signed in 2007. The interviews did not provide much insight into gender differences among youth with migrant backgrounds in Côte d'Ivoire.

Family rupture due to the death of one or both parents, or divorce and/or remarriage, can also bring about involuntary relocation for children and young people, as explained by a 17-year-old woman in Korhogo, northern Côte d'Ivoire, and a 24-year-old woman in Luweero, Uganda:

> My father did quite well, so we were fine until he died, then we began to suffer. As my grandmother was on her own, I was sent to live with her, and because I was helping her, I wasn't enrolled in school. When I turned 15, my family brought me to the village to get married. It was all very strange! There is no change in my everyday life. Before, when I was with my grandmother, I spent a little money on clothes and the rest was for our food, and now I give most of my money to my mother-in-law.

> My parents divorced and I stayed with my father on the farm. At 13 years old, I was brought to Kasubi to stay with my granny. This was because a worker at the farm was sexually harassing me, and staying at Kasubi saved me from this. The house was crowded with many people, and my needs could not be met, so I got myself a boyfriend and that was how I ended up pregnant at 15. It was all due to poverty because he had promised to provide for me. [...] I never knew my mother, but after struggling much in life, I felt I should find her. I was directed to where she was staying but she had married again and never told her husband that she had a daughter, so she introduced me to her husband as the daughter of her deceased sister. I felt so bad. I came to Nsozi Bilye in 2016 to take care of my sick granny and I ended up staying for good. I will have to migrate at some point in time, of that I am sure. I came to take care of my granny so if she happens to pass on, then I will have no reason to stay here. I doubt her children would let me stay even if I wanted to. They would have to divide up the land and I have no share here.

Even in situations of involuntary relocation young people find space to make choices, albeit from a limited range, which shape their transitions and social mobility. For girls, coming of age has a strong impact on their pathways depending on how sexuality and respectability is regarded in their community, and what age is common for intimate relationships and marriage. The reactions to having a marriage arranged or being pushed into deep poverty to be protected from unwanted sexual advances, as we see in the stories above, range from acquiescence to seeking a way out. Exiting a bad situation often involves relocation to the house of another relative or cohabiting with a boyfriend or prospective husband. Young men's choices rarely implicate life course markers such as marriage but rather livelihood enablers such as education and work. Nevertheless, consideration of family relationships and long-term access to resources may guide their choices of living with specific kin or discontinuing schooling.

Relocation for education

Chapter 7 (this volume) refers to the disruption in children's and young people's lives arising from relocation, following, for example, the death of a parent, parental divorce, 'fostering', etc., and the disjointed educational trajectories that can result. As the affected child or young person usually has little if any choice in the matter (i.e. whether and/or where to relocate), these cases are similar to, although possibly less dramatic than, some of the involuntary relocations discussed in the previous section. As with some of the other examples of involuntary relocation, while negative impacts on education are possible, this is not always the case:

> When I was about 2 years old, I was taken by my paternal uncle to live with him. I was told he wanted to relieve my parents from the responsibility since they didn't have money to efficiently care for me. He took me to Kizito-Luweero and I grew up there. I studied at Luweero Apex Primary School up to Primary 7 and my uncle paid for my school fees. I only came back to Kakuuto after my Primary 7 exams. I then went to Luweero Secondary School for ordinary and advanced level.
> (22-year-old female, Awach, Uganda)

However, the focus in this section is on relocation by young people when the primary motivation is to access a wider array or better quality of educational opportunities than those available locally. As should be expected – both because of the number and spatial distribution of different kinds of education establishments, and because young people's agency increases as they grow older – this kind of relocation is not very common among primary school students, but is increasingly common among secondary school, vocational and university students. It also figures in the futures that some young people imagine for themselves (see Chapter 9, this volume). In most cases, relocations for secondary or tertiary education involve, or will likely involve, young people moving from a rural area to a town or city. This is so common in Côte d'Ivoire that youth mention their guardians, but rarely detail how the arrangements came about or what their lives are (were) like, living in these households. Most mention that they help their guardian outside school hours. Some work to contribute towards their school expenses, while others explain that their guardian does not allow them to do work on their own account but wants them to focus on their studies. Most note that their school expenses are paid by their father, or sometimes their mother or a brother, often supplemented by monies they earn during weekends and holidays. A 20-year-old man in Côte d'Ivoire explains:

> I went to secondary school in Soubré and usually worked during the long vacations because I thought, if I don't do this, my dad can't send me to school. I did masonry and painting. I was forced to play truant from school to do piecemeal work, since my father didn't always send money. One time, I spent 2 weeks painting with my brother in Abidjan. I have also helped one of my older brothers to sell clothes in San Pedro, especially during the month of December. This year, when the new school year was about to start, my father told my siblings and I that he didn't have the money to send us to school. I was so discouraged that I took my stuff and went to my mother's house [his parents have divorced]. I only stayed for 2 weeks but when I came back, I learned that my younger brothers had started school, only I wasn't going back. Later I learned he [my father] had taken a credit to send them to school and I didn't want him to become more indebted, so I dropped out of school a year before I would have sat the BEPC [*Brevet d'Etudes du Premier Cycle*] exams.

A similar pattern is observed in Nigeria, where secondary and tertiary education involved mobilities of different durations:

> I left my father in 2015 at the age of 19 years. I left to pursue my education in Omu Aran [a location near Oshogbo in Osun state]. At that time, I came home every weekend to support my father. He was good to us and catered for us. When I finished my secondary level examination papers at Omu Aran, I did not come back to Oba Oke. I travelled to Offa [in Kwara state, a north central state in Nigeria], to pursue admission to Offa Polytechnic but also to participate in charcoal business with some friends. All the savings generated from these activities were used for clothing.
> (25-year-old male, Oba Oke, Nigeria)

In our research sites, the range of mobilities linked with young men's secondary education is infrequent among the young women. Those who reach secondary level education are more likely to work with their parents or guardians, to work within the locality where they live, or to come from families that can cover the whole cost for their education. Often young women are perceived to need more protection and they are able to protect their reputation and respectability by remaining in proximity of their kin. Their diligence in the economic activities they undertake factors in the level of maturity and social status they are accorded, even if they do not have the same level of geographical knowledge and breadth of social networks as their male peers.

Educational relocations are not just about access but also about the pursuit of a better or a different future. For example, in Uganda, a

29-year-old agronomist had laid the foundation for his future life by relocating from Luweero to Kampala to attend university, where he completed a first degree in agribusiness management. Perhaps reflecting his age, degree and his current position as an agronomist on a commercial farm, the future he imagines for himself is more expansive than many of the other interviewees. Already married with two children, he has established himself as an adult man and is now contemplating building his social status. He imagines himself getting a master's degree, moving to Busula (where he already has a plot of land), upgrading his political activity from a village councillor to a district councillor, and upgrading his wife's shop from retail to wholesale (she is also a university graduate).

A number of the interviewees in Ethiopia have completed a university degree, are currently studying for a degree, or hope to go to university and/or have siblings who have a degree. It is striking that in these interviews there was no mention of mobility: it seems to be simply assumed that relocation is nothing more or less than part of going to university (as it is in so many other places!).

Parents and other family relations generally supported mobility to access higher quality education, due in part to wanting their offspring or wards to succeed in life. The presence of close relatives, mainly paternal uncles living in areas with better schooling, is an important part of the story. Such relatives sometimes contribute financially to help defray the cost of formal schooling, and this, to a large extent, helps to facilitate mobility for education.

Vocational training was also part of young people's strategies, and for many, such training was a welcome solution if they had had to drop out of school. The ability to get involved in vocational training was highly gendered, and the transition to adulthood hampered some women's training because of childcare responsibilities and/or the demands of existing business activities:

> I would be interested in vocational training but not as part of formal education. I can do a more advanced training in [beauty] salon work and maybe add on with tailoring if there is money and time. Then, when I don't have customers for the salon, I can do the tailoring work. [...] The first major constraint is lack of money to pay for the vocational training and to buy the equipment. The second constraint is lack of time because this will require me to leave my home and maybe go to stay in Gulu, yet I have young children.
>
> (18-year-old female, Awach, Uganda)

In addition to childcare, other obstacles have stopped women from entering vocational training. A 33-year-old woman with two children from Butuntumula, Uganda, thought that vocational training was a possibility as it did not require relocation:

> I only want to go for vocational training to learn handwork like tailoring and running a [beauty] salon. There is a vocational institute nearby. People who train there start their own businesses. The major constraint is money. As for formal education, my husband may not allow it.

Access to vocational training is easier for unmarried women but the economic standing of the family might influence the kind of training, formal or informal, as may mediation by relatives in arranging a more formalized apprenticeship. How skills acquisition is labelled differs across the four countries. In Nigeria, young women who stopped schooling in junior or senior secondary school, talk about having trained as a tailor, seamstress, hairdresser or healthcare professional, regardless of whether they learn by working with a parent or in a vocational school. In contrast, a distinction is made in Côte d'Ivoire between apprenticeships that incur the payment of a fee and working alongside a relative. Young Ivorian men speak of being 'assistant' masons and painters, but 'apprentices' when they learn carpentry, welding and mechanics.

Such training can, but does not necessarily, involve relocation to other places, though the access to apprenticeships is generally easier in towns and cities. There is a strong sense among young people across the study sites that their rural areas lack opportunities for acquiring academic and practical skills, and therefore relocation must accompany any additional education (or training). For example, a young man in Idi Amu and Igbokiti, Nigeria, describes how there is no opportunity locally to be trained in the artisanal work he has chosen, so in order to 'realize his dream' he has to move to a bigger town:

> Like my own situation, I cannot learn my artisan job in this community so I chose to go out. This also applies to other young people who cannot be trained in their chosen work in this community.

Two other men from the same community voice a similar sentiment:

> I want to move out. [...] if you look around you won't see any higher institution here so this will make me move out and go to study.
>
> (18-year-old male, Nigeria)

> I can never further my education in this community. I prefer to move to a location like Sokoto [in northern Nigeria]. This community may distract someone from their original plans.
>
> (28-year-old male, Nigeria)

It has long been recognized that for some young people, relocation for education may be a first step in a longer journey, with further mobilities being facilitated by new skills and/or a broader perspective gained through education. This is clearly illustrated by a 22-year-old woman from Awach, Uganda:

> When it comes to [...] my university life, I will stay in Kampala. So, I am in Kampala basically for education since it has the best universities. After graduating, I feel I should explore another region of Uganda. I have friends from Adjuman and Arua so I plan to go there. Based on my friends, I think northerners are well-behaved people, kind and easy to get along with. I have started learning the language as a strategy to fit in the region. I don't think of going abroad at all. Who would take me there? I would be the first in my family to go abroad and it's kind of scary for me.

Relocations for work

Much attention has been paid in academia and policy to labour migration in SSA, especially to circular movements within countries and transnationally, and to irregular migration into Europe and North America. In fact, our research suggests that transnational migration is probably the least common pathway for rural youth. Only in Ethiopia was there evidence of a substantial flow of transnational migrants, reflecting the significant increase in the numbers of young women relocating to the Middle East for domestic work in the past 20 years (Atnafu and Adamek, 2016; Zewdu, 2018). The experience of an 18-year-old woman from Kuyu, Ethiopia, is not atypical:

> I interrupted my education halfway into Grade 9 to migrate to Bahrain for work, without telling my family. I went through other people. But as I did not pay money for going, when I arrived in Bahrain the cost of my travel was deduced from my salary. The broker's payment was also deduced, so for the first year I barely had any income and I did not have money to send to my mother. The work was very difficult for me. I had signed a 2-year contract, but I did not finish the 2 years. I stayed there for 1 year and 3 months and returned with 3 months' salary. I came back with the intention of changing my destination to another country, but my family prohibited me from migrating again. To some extent, I was able to furnish our house. I replaced the old furniture.

On the other hand, mobilities associated with transnational relocations are central in the imagined futures of some Nigerian youth:

> Oh, if God blesses me, the first place that is in my mind is America. It is a country that I love. When I go there with my music, I will go up [get famous] quick. I believe that about my music! You know, 3 years ago I met one man in Ife and he called me during the show, when I was performing there. He said that he liked the way I sing, and if I bring 50,000 Naira he will help me get the visa to go to Malaysia. I just tried my best and gave him the money but then later I did not see him again.
>
> (22-year-old male, Oba Oke, Nigeria)

While locally significant, these kinds of transnational labour movements do not seem to reflect the use or experience of mobility of the vast majority of rural youth. Their mobilities, associated with work and income generation, and with accruing social status through economic activities, are likely to be on more limited spatial and temporal scales. While they tend to be under the policy radar, the use of such mobilities by young people to shape and build their livelihoods deserve more fine-grained attention.

Labour migration research tends to home in on employment – in a traditional sense – as working for someone, but a focus on mobilities highlights the significance of even small-scale movement for trading-based economic activity.

A 29-year-old man in Butuntumula, Uganda, illustrates the centrality of such mobility to an (ultimately unsuccessful) trading venture:

> After dropping out of school in Primary 7, I started trading in fish. I bought fish from Luweero Town and cycled to the villages around Vumba to sell them. Being an amateur in business it collapsed due to bad debts.

Some nascent mobile traders have been fortunate to learn the ropes from a parent or others. In Côte d'Ivoire, for example, young men in particular become involved as mobile agents in agricultural value chains, working as microbuyers of cocoa and cashew nut in well-established commercial networks.

> When the cashew season begins, I go to the hamlets and small villages to buy. Some producers call me, and I go to their homes to buy. I work with a big trader in Korhogo, who tells me the price of cashew nut daily and buys all the cashew nuts I acquire. When the cashew season finishes, I move into trade in farming products. I started working with the trader during the time I was in secondary school. When I wasn't in school, I worked in his shop and he would give me 1–2000 FCFA.
>
> (28-year-old male, Korhogo, Côte d'Ivoire)

Such mobility-based trade in Côte d'Ivoire is highly gendered and provides a particular niche for young men, who combine it with a portfolio of farm and/or non-farm economic activities. They are well integrated into the rural environment and establish durable social networks through their different strands of work. Although they earn money and respect in their families for their work, these young men are still firmly embedded in a hierarchical social order in which they are subordinate to one or more larger (and generally older male) traders.

In Nigeria, a few young women provided examples of how their livelihoods depended on local mobilities for trade. For example, a 21-year-old woman in Oba Oke combines trade in agricultural produce with an itinerant hairdressing service:

> I engage in the buying and selling of agricultural produce. I buy raw cocoa beans, to dry and sell. I buy dried cocoa beans and resell as well. I go to nearby villages to buy cocoa and cashew nuts. When palm oil is harvested, I buy palm kernels to process and sell to retailers. I spend the income I generate for the upkeep of my family and myself. In the beginning my husband didn't know this job, and we always quarrelled when I came home late from buying and selling cocoa beans, kola and palm oil. But his mother intervened and explained that he had to be patient because the little income earned from the work I did would assist and improve the family. Now he permits me to continue my work. I always carry my tools for hairdressing when I travel so I also do hair and this links me with more people in the areas where I buy palm oil kernels in large quantities.

In contrast to the young men in Côte d'Ivoire who are integrated into established trading networks, this young Nigerian woman uses the female niche of hairdressing to build and strengthen her trading networks. Combining trade and hairdressing means her short-distance circular movements support a portfolio of income-generating activities consolidated through the time spent with other women. It was only with the support of her mother-in-law that she was able to negotiate space with her husband for her mobility. The mother-in-law's intervention suggests that these negotiations may have been much more difficult if her income was not being invested in her young family.

Examples of relocations to rural areas away from the villages we studied, emerged primarily in conversations about young people's plans for the future. In Uganda, land markets and the possibilities they allow for buying or otherwise accessing land in other parts of the country, were important for young married men who were struggling to provide for their families, and thus to live up to patriarchal expectations:

> I want to buy more land away from home and I have been saving since last year. The place where I want to acquire the land is in Amuru District where the land is very fertile. My plan is to buy land for farming and for constructing a house. I also want to buy more animals. I will migrate when I get the money for buying the land and means of transport for easy movement to the farm.
>
> (32-year-old male, Awach, Uganda)

Land scarcity is only one among several drivers of mobility, which tend to play out through gender divisions of labour. For example, in areas where women do not participate much in farm work or do not have their own farms, relocation for work seems more acceptable, at least as long as the women are single. For married women, relocation requires agreement with the husband to move alone or together.

In sites across all four countries, some young people reported either having relocated, or were planning to relocate to rural towns for shorter or longer periods of time. Young women in particular spoke of their wish to relocate to rural towns in the future to expand their businesses as tailors, hairdressers, shop owners, restaurant owners, etc.:

> If it is the will of God, I will move to a town to engage in different types of work. I want to live in a town and engage in any activity that can generate income there. For example, I want to open a cafeteria. If my capacity does not allow me, I would open a shop or engage in other trading activities. Rural work is not convenient for me, especially because of my health, but I also think that the activities in town are more profitable than activities in the village.
>
> (26-year-old female, Jabi Tenan, Ethiopia)

Young people's mobilities do not necessarily follow a pattern from periphery to centre, or from lowly- to better-paid work. Indeed, the step from rural towns to more densely populated urban areas is often insignificant and more dependent on the distance from village to city, the availability of work and the ability to save up money, than on the size of the population. In many cases the shift to a new place appeared to be more random than planned:

> A friend helped me to get a job as a maid in Entebbe and I was there for a year. My salary was never paid on time, and there was a time when they owed me 200,000 Shillings but paid me only 50,000. When I told my employer that I had got a new job in Kikyusa, she took me to the taxi park and told me to wait while she got the money she owed me from her mobile money. But she never came back. Another friend helped me get the job in Kikyusa, working as a maid for a Rwandese woman. I worked in this woman's home for close to 2 years and it was there that I met and married my husband.
>
> (25-year-old female, Luweero, Uganda)

The story of unpaid wages is common, and particularly among young people who have relocated to places where they do not have older members of their family to intervene with the employer. Peers cannot play the same role as older kin, but they can help in facilitating alternative employment, which may require mobility within the city or rural town. When young women and men voice their wish to migrate for urban work in the future they tend to speak in terms of ambitious goals and to glorify the potential outcomes. The same type of embellishment happens in discourses about Europe among our Ivorian respondents. A 29-year-old widow and mother of four in Odienné, Côte d'Ivoire, notes with a laugh:

> My wish is to go beyond Abidjan, but it is really difficult to know where it is best to be. If you are in a place and everything goes well for you, it is clear that it's better, but if you sit somewhere and you don't have anything, then you are not better off than you were before.

Some of the dreams about distant places seem unrealistic, and even some plans for more local migration appear lofty and barely attainable. However, this amalgam of dreams and imagined futures must be understood as a reflection of the diverse livelihood activities in which young people engage, and their familiarity with hazard and contingencies that can easily turn their lives upside down.

Discussion and Conclusions

This chapter has explored young people's mobility and highlighted gender differences that shape what young rural women and men do, and how this is interlinked with transitions and social mobility. Qualitative data from Uganda, Ethiopia, Nigeria and Côte d'Ivoire show that the historical and continued importance of mobility in much of rural Africa does not translate into rural exodus, but rather into changing patterns of mobility. By shifting the focus from migration to mobilities, the chapter overcomes disparities in what is defined as migration, what are understood as ordinary relocations (for example for education or marriage), and what can be seen as strategic mobility to access resources. We have not discussed practices that require women to move to their husbands' households, which are dominant across our research sites, in terms of how they affect women's access to resources. But we do highlight how norms about gender and age-appropriate behaviour, and to some extent class, affect women's ability to use strategic mobilities to access opportunities, pursue the options that they would prefer, and grow their economic activities.

Ambiguities in the distinction between voluntary and involuntary mobilities surface throughout the chapter, and reveal that even when relocation is prompted by violence or rupture, some choices are still made by adults and young people alike. Relocations are plural, flexible and repeated, especially in protracted conflicts, and they have different long-term outcomes for young women and men. These differences are rooted in gendered practices in relation to land tenure, but also to perceived needs for protection against abduction, brutality and sexual violence. Protective measures may involve relocation to a relative's household, but our analysis shows that young women have some agency to navigate difficult or unwanted relocations. Sometimes they do this by relocating on their own initiative, and sometimes by engaging in intimate relationships, and gaining the social status that comes with being married and having children. Young men are not often pressured into relocations that catapult them into life course transitions, but involuntary relocations during protracted conflict can affect their ability to pursue educational or occupational paths to more secure livelihoods. And yet, our analysis shows that multiple relocations may also be part of accessing educational or occupational opportunities.

Relocation for education is not limited to conflict zones or cases of familial rupture: indeed, in the majority of our research sites such mobility is taken for granted. For secondary education, relocations are often of shorter distance but may nevertheless involve living away from the home village. Tertiary education almost always requires relocation over longer distances. While the perceived quality of education enters into the decisions about which school to choose, our analysis shows that for secondary education the social relations with potential guardians, and the availability of subsidized school attendance, or of work opportunities, influence both the choice of institution and educational outcomes. The analysis also shows the importance of mobility in vocational training, and the constraints facing young women who have already transitioned into adulthood as wives and mothers.

Relocation for work is the type of mobility that has received the most attention; framed as 'circular labour migration' and increasingly just as 'labour migration', the implication is that migrants will settle elsewhere and not return. However, these framings ignore the contingent nature of young people's mobilities and livelihoods. Our analysis shines a light on how students' journeys during the long holidays can turn into temporary work and vocational training, and allow young people to occupy an age-appropriate space, and reconstruct age-appropriate relationships, within the family.

Policy discourses tend to use crude distinctions between different types of mobility – such as rural-to-rural versus rural-to-urban migration, or internal versus international migration – which do not adequately reflect the multiple mobilities observed, nor the differences associated with gendered norms and practices. Apart from the established flow of female temporary migrants from Ethiopia to work in domestic service in the Middle East, few young people are able to – or seemingly particularly interested in – relocating outside SSA. Nevertheless, the majority of young women and men are already engaged in, or envisage, strategies for broadening and consolidating their economic activities and achieving social adulthood, involving mobility in profound ways.

References

Atnafu, A. & Adamek, M. E. 2016. The return migration experiences of Ethiopian women trafficked to Bahrain: '... for richer or poorer, let me be on the hands of my people ...'. *African and Black Diaspora: An International Journal,* 9, 243–256.

Benwell, M. C. 2009. Challenging minority world privilege: Children's outdoor mobilities in post-apartheid South Africa. *Mobilities,* 4, 77–101.

Both, J. 2010. Peer groups and human anchorage: Girl migrants making it work in NDjamena, Chad. *In:* De Bruijn, M. & Merolla, D. (eds.) *Researching Africa: Explorations of Everyday African Encounters. African Studies Centre, African Studies Collection, Vol. 26.* Leiden, Netherlands, African Studies Centre.

Camfield, L. & Tafere, Y. 2009. 'Children with a good life *have* to have school bags': *Diverse Understandings of Well-being Aiming at Older Children in Three Ethiopian Communities*, Oxford, UK, Young Lives.

Carling, J. & Schewel, K. 2018. Revisiting aspiration and ability in international migration. *Journal of Ethnic and Migration Studies,* 44, 945–963.

Castle, S. & Diarra, A. 2003. *The International Migration of Young Malians: Tradition, Necessity or Rite of Passage?,* London, London School of Hygiene & Tropical Medicine.

Cole, J. 2010. *Sex and Salvation. Imagining the Future in Madagascar,* Chicago and London, The University of Chicago Press.

Crivello, G. 2011. 'Becoming somebody': youth transitions through education and migration in Peru. *Journal of Youth Studies,* 14, 395–411.

Crivello, G. & Van Der Gaag, N. 2016. Between hope and a hard place: Boys and young men negotiating gender, poverty and social worth in Ethiopia. Oxford, UK, Young Lives.

De Bruijn, M., Van Dijk, R. & Foeken, D. (eds.) 2001. *Mobile Africa: Changing Patterns of Movement in Africa and Beyond,* Leiden, Netherlands, Brill.

Farrugia, D. 2016. The mobility imperative for rural youth: the structural, symbolic and non-representational dimensions rural youth mobilities. *Journal of Youth Studies,* 19, 836–851.

Gaibazzi, P. 2018. *Bush Bound. Young Men and Rural Permanence in Migrant West Africa,* New York and Oxford, Berhahn.

Gough, K. V. 2008. 'Moving around': the social and spatial mobility of youth in Lusaka. *Geografiska Annaler: Series B, Human Geography,* 90, 243–255.

Gough, K. V. & Birch-Thomsen, T. 2016. Mobile rural youth in northern Ghana: Combining near and distant opportunity spaces. *In:* Gough, K. V. & Langevang, T. (eds.) *Young Entrepreneurs in Sub-Saharan Africa.* Abingdon, UK, Routledge.

Hannam, K., Sheller, M. & Urry, J. 2006. Editorial: Mobilities, immobilities and moorings. *Mobilities,* 1, 1–22.

Hashim, I. 2005. *Exploring the Linkages Between Children's Independent Migration and Education: Evidence from Ghana,* Brighton, UK, Sussex Centre for Migration Research, University of Sussex.

Hashim, I. & Thorsen, D. 2011. *Child Migration in Africa,* New York, Zed Books.

Hashim, I. M. 2007. Independent child migration and education in Ghana. *Development and Change,* 38, 911–931.

Kleist, N. & Thorsen, D. (eds.) 2017. *Hope and Uncertainty in Contemporary African migration,* New York and London, Taylor & Francis.

Langevang, T. & Gough, K. V. 2009. Surviving through movement: the mobility of urban youth in Ghana. *Social and Cultural Geography,* 10, 741–756.

Meinert, L. 2003. Sweet and bitter places: the politics of schoolchildren's orientation in rural Uganda. *In:* Olwig, K. F. & Gulløv, E. (eds.) *Children's Places. Cross-cultural Perspectives,* London, Routledge.

Porter, G., Blaufuss, K. & Acheampong, F. 2008. Youth, mobility and rural livelihoods in sub-Saharan Africa: Perspectives from Ghana and Nigeria. *Africa Insight,* 37, 420–431.

Porter, G., Hampshire, K., Mashiri, M., Dube, S. & Maponya, G. 2010. 'Youthscapes' and escapes in rural Africa: Education, mobility and livelihood trajectories for young people in Eastern Cape, South Africa. *Journal of International Development,* 22, 1090–1101.

Porter, G., Hampshire, K., Munthali, A. & Robson, E. 2011. Mobility, surveillance and control of children and young people in the everyday: perspectives from sub-Saharan Africa. *Surveillance & Society,* 9.

Porter, G., Hampshire, K., Abane, A., Munthali, A., Robson, E. & Mashiri, M. 2017. *Young People's Daily Mobilities in Sub-Saharan Africa: Moving Young Lives,* New York, Palgrave Macmillan.

Punch, S. 2014. Young migrant trajectories from Bolivia to Argentina: Changes and continuities in an era of globalisation. *In:* Veale, A. & Dona, G. (eds.) *Child and Youth Mobility in Migration: Mobility in Migration in an Era of Globalization,* Basingstoke, UK, Palgrave Macmillan.

Punch, S. 2015. Youth transitions and migration: negotiated and constrained interdependencies within and across generations. *Journal of Youth Studies,* 18, 266–276.

Tetteh, P. M. 2014. Child domestic labour in Accra: A juxtaposition of the myths with the reality. *Childhoods Today,* 8.

Thorsen, D. 2006. Child migrants in transit. Strategies to become adult in rural Burkina Faso. *In:* Christiansen, C., Utas, M. & Vigh, H. E. (eds.) *Navigating Youth, Generating Adulthood: Social Becoming in an African Context,* Uppsala, Sweden, Nordic Africa Institute.

Thorsen, D. 2012. *Children Working in the Urban Informal Economy: Evidence from West and Central Africa. UNICEF Briefing Paper Number 3*, UNICEF.

Thorsen, D. 2014. Jeans, bicycles and mobile phones. Adolescent migrants' material consumption in Burkina Faso. *In:* Veale, A. & Donà, G. (eds.) *Child and Youth Migration. Mobility-in-migration in an Era of Globalization*, Basingstoke, UK, Palgrave MacMillan.

Thorsen, D. & Jacquemin, M. 2015. Temporalités, savoir-faire et modes d'action des enfants travailleurs migrants au sein de la parenté élargie en Afrique de l'Ouest. *Canadian Journal of African Studies/ La Revue Canadienne des Etudes Africaines,* 49, 285–299.

Veale, A. & Donà, G. 2014. Complex migrations, migrant child and family life trajectories and globalization. *In:* Veale, A. & Donà, G. (eds.) *Child and Youth Migration. Mobility-in-migration in an Era of Globalization*, Basingstoke, UK, Palgrave MacMillan.

Yeboah, T. 2020. Future aspirations of rural-urban young migrants in Accra, Ghana. *Children's Geographies*, 1–14.

Zewdu, G. A. 2018. Ethiopian female domestic labour migration to the Middle East: patterns, trends, and drivers. *African and Black Diaspora: An International Journal,* 11, 6–19.

Zongo, M. 2016. La diaspora Burkinabè en Côte d'Ivoire : entre vulnérabilité et quête d'appartenance. *In:* Bredeloup, S. & Zongo, M. (eds.) *Repenser les Mobilités Burkinabè,* Paris, L'Harmattan.

6 Are Young People Transforming the Rural Economy?

Jordan Chamberlin[1] and James Sumberg[2]
[1]*International Maize and Wheat Improvement Centre (CIMMYT), Nairobi, Kenya;*
[2]*Institute of Development Studies, Brighton, UK*

Introduction

> Youth are at the forefront of championing the innovative technological, gender-aware, and climate-smart approaches that will help grow and modernize agriculture.
>
> (Young Africa Works Summit 2017, MasterCard Foundation)[i]

> When the willingness of youth to contribute is matched with appropriate opportunities, they can have a transformative impact on the growth of agricultural productivity.
>
> (Alliance for Green Revolution in Africa (AGRA), 2015, p.38)

In recent years, it has often been asserted that young people are innovative, risk-taking, early adopters of new technologies, and eager to engage with non-traditional opportunities. Within the context of sub-Saharan Africa (SSA), these putative essential features of rural young people have been proposed as a potential driver of change in agriculture and other parts of the rural economy (Africa Commission, 2009; FAO *et al.*, 2014; AGRA, 2015). However, the argument continues: resource constraints and limited productive opportunities in SSA's rural areas limit the ability of youth to realize their potential as farmers, and increase the likelihood that talented young people abandon rural areas altogether through migration (see Chapters 3, 5 and 9, this volume). In this sense, rural-to-urban migration, as a conduit for rural youth away from agriculture, is often framed as having potential negative impacts on the agricultural sector and rural development (The World Bank, 2015; Deotti and Estruch, 2016).[ii] This chapter examines the empirical basis for this story, using a combination of approaches.

The chapter sets out to address three questions. The first is, 'How might we think about the notion that youth bring something new to farming?' To address this, we examine the explicit and implicit assertions in the stylized statements which are often made in policy and public discourse about young people and farming in SSA. This is partly a conceptual parsing of the different ways in which we might interpret claims about youth and how they intersect. We complement this with a review of evidence about particular propositions (youth are innovative; youth will drive agricultural transformation) that appear in the literature, drawing on other recent studies.

Second, we address the question of 'What aspects of farming by young people are visible to us through existing empirical windows?' Not all implied traits of youth as farmers are equally visible in household survey data. For example, the fact that agricultural decisions and outcomes are typically described at the farm or household level imposes challenges for linking such outcomes with youth, an attribute generally ascribed either to individuals within a household or to a whole demographic segment. We discuss the opportunities

DOI: 10.1079/9781789245011.0006

and constraints associated with using currently available data sources to explore how young people farm in Africa.

Third, given the caveats outlined in addressing the preceding questions, and conditional on the data that are currently available, we ask, 'Do the young in SSA farm differently?' We use recent nationally representative household survey data to address this question, focusing on observable farm-level practices which are generally linked with agricultural intensification. We find that younger households (whether evaluated by age of household head or average age of household adults) are characterized by structural differences, typically operating smaller farms with fewer capital resources. However, farms operated by younger households are also slightly more likely to use intensification practices and be oriented toward markets. After controlling for farm endowments and other factors, we find evidence that younger households are significantly more likely to use intensification practices: age of head and age of workforce are both negatively associated with output marketing, high-value/non-traditional production portfolios and input usage. Interestingly, after controlling for age of the household head, we find a significant additional effect of the age of other household members, suggesting that relying exclusively on characteristics of the head as 'farm manager' may obscure important contributions made by other young household members to decision making. However, while statistically significant, our results indicate that the 'age effect' in the use of intensification practices is very small in magnitude, implying that expectations of a youth-driven transformation of African agriculture should be examined carefully.

Change in Rural Economies

Young people as agents of change

The idea that African young people can be the agents for positive change in rural areas is in wide circulation. For some, they are viewed as the only hope for achieving sustainable agricultural transformation, while for others, their positive influence is already being felt. These claims have been critically reviewed elsewhere and found wanting (Mabiso and Benfica, 2019; Sumberg and Hunt, 2019; Glover and Sumberg, 2020) and we see no need to cover this same ground again here. Suffice it to say that there is little if any evidence – direct or indirect – from rural SSA that supports these kinds of propositions. Further, as shown by Sumberg and Hunt (2019),

> […] The international research literature on the relationship between age and a number of indicators of creative and innovative behaviour, point to the conclusion that there is no clear or strong evidence for a simple and/or direct relationship. Further, in those individual studies where such relations are found, they are often influenced by context and sector, and individuals at the lower end of the youth age range are seldom the most innovative or creative.
>
> (Sumberg and Hunt, 2019, p.135)

In addition to the lack of evidence, the claims that young people are in pole position to transform African rural economies are undermined by three critical weaknesses. The first is the fact that they fall right into the 'essentialism trap' by suggesting that young people, as a group, are hungry for change, and are innovative, quick to adopt new technology, etc. This kind of conception of youth as a uniform, undifferentiated group – across factors of social difference including gender, age, ethnicity, religion, level of formal schooling, wealth, etc. – flies in the face of the basic tenets of modern sociology. It has taken decades to break down entrenched mentalities behind statements like 'women are …', and it would be a shame if similar essentialisms are allowed take root in the discussion of rural youth.

The second weakness is muddled thinking, in which innovation and innovative behaviour are used interchangeably with, for example, trying new things, risk taking and entrepreneurship. The danger in using quite different ideas such as these, as synonymous, is that we run the risk of devaluing important, long-standing analytical distinctions. This is clearly seen, for example, in the case of entrepreneurship, which in current usage in relation to youth in SSA – following the approach of the Global Entrepreneurship Monitor[iii] – is now often undifferentiated from any kind of self-employment (Kew *et al.*, 2015).[iv] Gough and Langevang (2017) capture this well: 'In light of their limited possibilities to gain formal sector jobs in the public or private sector, young people are being encouraged to be "job creators" rather than "job seekers", thus becoming self-employed "entrepreneurs"' (p.1).

The third weakness stems from the fact that claims about young people's potential to transform the rural economy are seldom if ever accompanied by an acknowledgement (much less any analysis) of the structural and political economy impediments that constrain their ability to drive change. In effect, the suggestion is that young people will drive system-level, transformational change, despite being young and inexperienced, without access to the full suite of resources required for meaningful innovation, and often excluded from the fora and institutions where key decisions are made.

Farming: the challenge of seeing the innovation through the difference

The notion of a farming system or a crop production system suggests it is possible and useful to categorize farming activity based on key indicators like scale, crop and/or livestock combination, commercial orientation[v], labour arrangements and the technology used. The objective in developing such system classifications is usually to maximize variation between, while minimizing it within, the set of identified systems. While the expectation is that all farms within a given category are broadly similar, there will always be a degree of variation in structure, organization, the use of technology, productivity and the like, within any given category. Some of this within category (or within system) variation reflects different resource endowments and management preferences and capabilities: it would certainly be a mistake to read within category variation as direct evidence of innovation or innovative behaviour on the part of an individual or household.

The concept of innovation – simply put, 'improved products and processes' (Niosi *et al.*, 1993) – has been central to the mainstream understanding of economic growth for many decades (Schumpeter, 1947). None the less, perspectives on the relationship between African smallholder farmers and innovation have shifted dramatically since the colonial period. Initially, with farmers portrayed as backward, tradition bound and resistant to change (for more sympathetic early portrayals, see Hill, 1963; Allan, 1965), it was the job of agricultural extension to introduce and promote modern ('rational') farming methods. Framed by diffusion of innovation theory, those who first adopted the new technologies and practices, i.e. they did what was being promoted, were termed 'innovators' (Rogers, 1983). It was not until some decades later that the existence and potential significance of farmers' own experimentation and innovation came briefly to the fore (e.g. Richards, 1986; Biggs, 1990; Sumberg and Okali, 1997).

Innovation is commonly used to refer to both the process through which new products and processes emerge, as well as the new products and processes themselves. The former is difficult to observe, particularly at farm level, while the use and spread of new products (like crop varieties) or processes (like conservation agriculture), promoted by extension or development programmes, are usually the focus of technology diffusion studies (Glover *et al.*, 2019; Krishna *et al.*, 2020).

For the purposes of this chapter it is useful to think of innovation or innovative behaviour in relation to the rural economy – or to young people's livelihoods – in terms of *doing different things* and/or *doing things differently*. A farmer who uses (not just tries) fertilizer when no one else does, might be considered innovative, as might one who applies fertilizer using a micro-dosing technique, when everyone else broadcasts. In addition to farm practices, innovation – doing different things and/or doing things differently – might be observed in the marketing or commercialization of agricultural produce; in farm structure or organization; in how agriculture is combined with other economic activities; or in any aspects of non-farm economic activities.

In this chapter we are interested in generational differences in farming practice, but the danger of simply ascribing such differences to innovation or innovative behaviour can be illustrated with the example of young tomato growers in Brong Ahafo, Ghana (Sumberg and Okali, 2006; Okali and Sumberg, 2012). In the study site north of Techiman, tomatoes – up to three crops per year – are produced, as they have been for some years, by young people (primarily but not uniquely young men). Plots are relatively small and the work is hard, particularly during the dry season when watering is required morning and evening. Many of the young tomato growers do not bother to grow food crops – having little if any household provision responsibilities – but focus solely on tomatoes because of the high potential return. Many rely on a mother or sister to feed them. The point is that the differences between the young tomato growers and the older generation, who grow a range of food crops and are unlikely to produce tomatoes, has little to do with innovation, and everything to so with the young people's physical strength, limited responsibility and desire to accumulate capital quickly. There is much more scope for looking at innovative behaviour among the tomato producers, where, for example, only a few make use of petrol pumps to facilitate irrigation or try to circumvent the women who control access to the market (Okali and Sumberg, 2012).

Evidence from technology adoption studies

The vast empirical literature on technology use offers decidedly mixed evidence about young farmers' inherent propensities to use modern intensification strategies. In a systematic literature review of the determinants of using conservation agriculture practices, Knowler and Bradshaw (2007) found that of 18 studies controlling for age of farmer, three found a negative relationship (i.e. younger farmers are more likely to adopt such strategies), five found a positive relationship, and ten found no statistically significant relationship. Their review of the literature also indicated no consistent finding with respect to farmer age and propensity for technology adoption. Meijer *et al.*'s (2015) literature review of factors affecting the uptake of agricultural and agroforestry innovations among smallholder farmers in SSA found no consistent connection between uptake and age of farmer. Kassie *et al.* (2015) found that after controlling for other factors, there was often a negative correlation between farmer age and adoption of different conservation agriculture practices in Kenya, Malawi, Ethiopia and Tanzania, although coefficient estimates were highly variable across practices and countries, and not always significant. Other recent empirical efforts find little evidence for a 'youth effect' in agricultural intensification. Diao *et al.* (2019) find that younger farmers in Ghana are not more likely to intensify or use modern inputs, compared with older farmers; constraints to adoption seem non-age specific (e.g. market access is important for technology adoption). Andersson Djurfeldt *et al.*'s (2019) mixed methods study in Zambia also found no age-related intensification or productivity relationship, which they attribute in part to the limited resources of younger household heads. In contrast, Guo *et al.*'s (2020) meta-analysis of 33 empirical studies found a significantly positive association between age and use of sustainable intensification practices in Southern Africa.

Empirical Windows on Young Farmers: Data and Measurement Challenges

In order to begin to evaluate some of the claims made – however vaguely – about young farmers and their potential contributions to the transformation of agriculture, we would ideally like to draw on well-measured observations of such characteristics as energy, risk aversion, innovation, receptivity to new ideas, early adoption of new technologies and so on. Standard household survey instruments, however, generally afford few windows on to these characteristics and the outcomes potentially associated with them.

The propensity to innovate is often ascribed to young people as a key characteristic.[vi] Innovation (like decision making) is not easily observable, almost by definition. Survey instruments are generally set up to track practices that are already 'on the radar', and predefined questions about farming practices will certainly miss innovation and experimentation at the margins. However, through these surveys we can generally observe a suite of intensification practices which are heavily promoted by extension and development programmes, including, for example, the use of improved crop varieties, inorganic fertilizer and other inputs. Increasingly, surveys are also tracking practices that map on to more agroecological or 'climate-smart' approaches to agriculture (Hammond *et al.*, 2017; Ngoma *et al.*, 2020).[vii] So, a first approach to evaluating the propensity of young farmers to drive change might examine the use of these two kinds of practices.

A number of challenges arise in attempting to link individual attributes (age, gender, education level or any other property of youth) with practices and outcomes typically observed at the farm level (e.g. technology use, productivity, farm orientation). First, many outcomes are only reported at the household level. Second, while it is true that in some cases we observe plot-level practices and outcomes, which in theory could be linked with different 'managers' within the household, in practice, this does not generate much analytical purchase. This is because in the Living Standards Measurement Study – Integrated Surveys on Agriculture data and similar data sets where plot-specific control and/or management is mapped to particular household members, the vast majority of plots are said to be managed by the head, with the majority of the remainder mapped to the head's spouse(s). For example, using nationally representative samples, 90% of plots in Zambia were managed by the head, 8% by the spouse and just 2% by other members, while in Tanzania and Ethiopia, the head was identified as the primary decision maker on 94% and 95% of cropped plots, respectively.[viii]

In recognition of the limitations of assigning a single household member as plot manager, recent surveys are increasingly asking about multiple contributors to decision making. Table 6.1 shows the share of cropped plots with decisions made by different combinations of household members, when joint or consultative decision making is recorded. While significantly more inclusive than descriptive statistics derived from a single manager, we still see relatively low incidence of members other than the head or spouse(s) recorded as contributing to decision making, with less than a tenth of Tanzanian plots and less than a quarter of Ethiopian plots having recorded decision making input from members other than the head and/or spouse(s).

This regularity in farm survey data has several possible implications. First, taken at face value, it suggests that for the most part young people do not farm (other than by supplying labour) until they become household heads or spouses. A second possibility is that young people are engaged in managerial decision making on at least some plots, either solely or with other household members, but this managerial contribution remains opaque given a default preference to designate the head as plot manager. There are many reasons to suspect that the second of these alternatives is likely in many cases. Consider the not atypical case of a household with an aged patriarch, his wife or wives, and several adult children who are still at home. Tradition may dictate that the patriarch is designated as the household head, even where the de facto management of many of the household's productive activities (on and off the farm) are in the hands of other members. To adequately explore this possibility, new modes of data collection will need to be developed that more carefully probe the locus (or loci) of decision making within

Table 6.1. Share of cropped plots by household position of decision maker (including joint or consultative decision making).

Decision maker	Tanzania	Ethiopia
Head only	43%	30%
Spouse only	3%	1%
Head and spouse	45%	45%
Other household member, alone or with another member	9%	24%
Total	100%	100%

Notes: Data are from the 2012–2013 Tanzanian Living Standards Measurement Study – Integrated Surveys on Agriculture (LSMS–ISA), and 2015–2016 Ethiopian LSMS–ISA surveys. For Tanzania, up to three household members are identified in response to the question 'Who decided what to plant on this plot in the long rainy season?' asked for cropped plots (up to three members were possible joint decision makers). For Ethiopia, for cropped plots, in addition to the to the question 'Who in the household makes primary decisions concerning crops to be planted, input use and the timing of cropping activities on this field?', up to two other members could be identified as consulted by the primary decision maker.

households. In the meantime, we might creatively reinterpret 'farmer age' (or other farmer characteristics) as something better measured through an aggregate measure of the family farm workforce. We offer an example of one way to undertake this in the next section. A third possibility, which is not mutually exclusive of the first two, is that young household members who are not the head or a spouse are farming on plots which do not show up in the household plot roster. To keep things tractable, our discussion in this chapter focuses on the first two alternative readings of the data. However, an important avenue for future empirical work would be to validate the assumption that information on all plots farmed by all household members is effectively captured by typical survey instruments.

Do the Young Farm Differently? Available Empirical Evidence

Despite these acknowledged limitations, we use data from nationally representative surveys of rural farm households in Tanzania, Zambia and Ethiopia to explore the degree to which alternative measures of household age correspond to multiple agricultural practices and outcomes. First, we describe the association with farmer age and structural characteristics of farms, such as size of landholdings and value of productive assets. This allows us to engage with debates on whether young farmers consistently face constraints in accessing land and other productive resources. Then, after controlling for these factors, we explore whether younger farmers are more likely to engage in intensification, agroecological or climate-smart practices.

The data we use for this analysis come primarily from the LSMS–ISA for Ethiopia and Tanzania, and the Rural Agricultural Livelihoods Survey (RALS) (CSO, 2012) for Zambia.[ix] All are nationally representative panel surveys, in which the same households were visited during the different waves of data collection. We use three waves of panel data for Ethiopia (2011–2012, 2013–2014 and 2015–2016) and Tanzania (2008–2009, 2010–2011 and 2012–2013) and two waves of data for Zambia (2011–2012 and 2014–2015). In Ethiopia, 3969, 3776 and 4951 households appear in each respective wave and in Tanzania, 3265, 3918 and 5010 households appear in each respective wave. In Zambia, 8839 households were surveyed in 2012, with 7254 reinterviewed in 2015.

To address the issue raised in the previous section about a potential mismatch between the household head and the locus of managerial agency in a distributed or collective set of decisions, we define several ways of measuring 'household age': age of the nominal household head; average age of all adult members; and average age of all adult members excluding the household head. The correlation between these measures is shown in Table 6.2, with very similar estimates for each country. We see that households with younger heads generally also

Table 6.2. Correlation between alternative measures of 'household age'.

Variables	Age of head	Avg. adult age	Avg. adult age (excluding head)
Tanzania			
Age of head	1		
Avg. adult age	0.7287***	1	
Avg. adult age (no head)	0.4056***	0. 8832***	1
Zambia			
Age of head	1		
Avg. adult age	0.7296***	1	
Avg. adult age (no head)	0.4650***	0.9056***	1
Ethiopia			
Age of head	1		
Avg. adult age	0. 6959***	1	
Avg. adult age (no head)	0. 2820***	0. 8367***	1

Notes: All Pearson correlation coefficients are significant at p<0.001 level (***).

have younger adult members, although this correlation is imperfect, suggesting that there may be value in evaluating alternative measures of household youthfulness.

We explore the proposition that younger farmers and households face particularly acute asset endowments, by plotting non-parametric associations between structural characteristics of farms and the age of the household. Fig. 6.1 shows these relationships graphically for Tanzania, Zambia and Ethiopia (represented by panels a, b and c, respectively). Here we use age of household head as our measure of household age, but the use of average adult age yields very similar results. For all three countries, farm size, labour and capital endowments are generally increasing with household age. Vertical lines indicate 25th, 50th and 75th percentiles of household head age distributions. The positive relationship between per capita land endowments and household age indicates that as households become older, landholdings are generally increasing faster than the increases in household size. The value of productive assets (only available for Tanzania and Zambia) are also increasing with age. By focusing on the left-hand side of these graphs, we generally find descriptive support for the idea that younger households operate with fewer productive resources than do older households, on average. Interestingly, the graphs of value of crop production per hectare (a frequently used measure of farm productivity) generally show an inverse U-shape, indicating that while resources tend to increase throughout the range of household ages, land productivity follows a different pattern, first rising and then falling (although peaking at different ages in each of the countries).

To further examine some of these structural conditions, as well as the practices and outcomes of interest, we tabulate descriptive statistics covering a broad range of characteristics on the samples from the most recent wave for each country, and use quartiles of household average adult age to organize our findings (Table 6.3). Several things stand out from these unconditional descriptive statistics. The number of members is greatest in the second quartile, and is consistent with an inverse U-shaped labour endowment which is lowest for very young and very old households. Farm size, farm size per household member and productive assets all increase with age. Echoing the findings in the graphs previously shown, we see that land endowments increase faster with age than labour endowments, meaning a growth in per capita land availability with age. The value of productive assets is highest in the middle range for Tanzania and increasing for Zambia (again, this variable is not available for Ethiopia). This is consistent with the stylized notion that many young households in rural Africa – and throughout the world – have relatively smaller asset endowments than older households.

In terms of intensification practices and outcomes, technology usage is highest generally in the second quartile, rather than in the youngest quartile. Overall, management practices do not vary strongly by age category. Farm orientation does not change dramatically across age categories, although the youngest households market a marginally higher share of their production. The value of crop production per hectare is declining across all categories for Zambia and Ethiopia, but is highest in the middle ranges for Tanzania. These unconditional results indicate that the youngest households are not obviously at the forefront of agricultural change. This could reflect lower resource endowments (land, labour, capita), but these young households are likely to be less experienced and may, for example, have their eyes on activities other than crop production.

To examine the relationship between household age and intensification practices and outcomes, we specify a number of regression models which control for resource endowments and other observable characteristics (Table 6.4). The practices of interest in these regressions are fertilizer usage, pesticide usage, herbicide usage and usage of improved maize seed, while the outcomes of interest are commercialized share of value of production and high-value share of value of production.[x] For ease of exposition, we show only the coefficient of interest from model specifications that control for age of head (rows labelled 'a'), the average age of household adults (rows labelled 'b'), and with simultaneous controls for age of head and average age of non-head members (rows labelled 'c'). Other control variables include the sex of household head, farm size, number of household members, the value of productive assets, distance from the nearest town of 50,000 or more inhabitants,

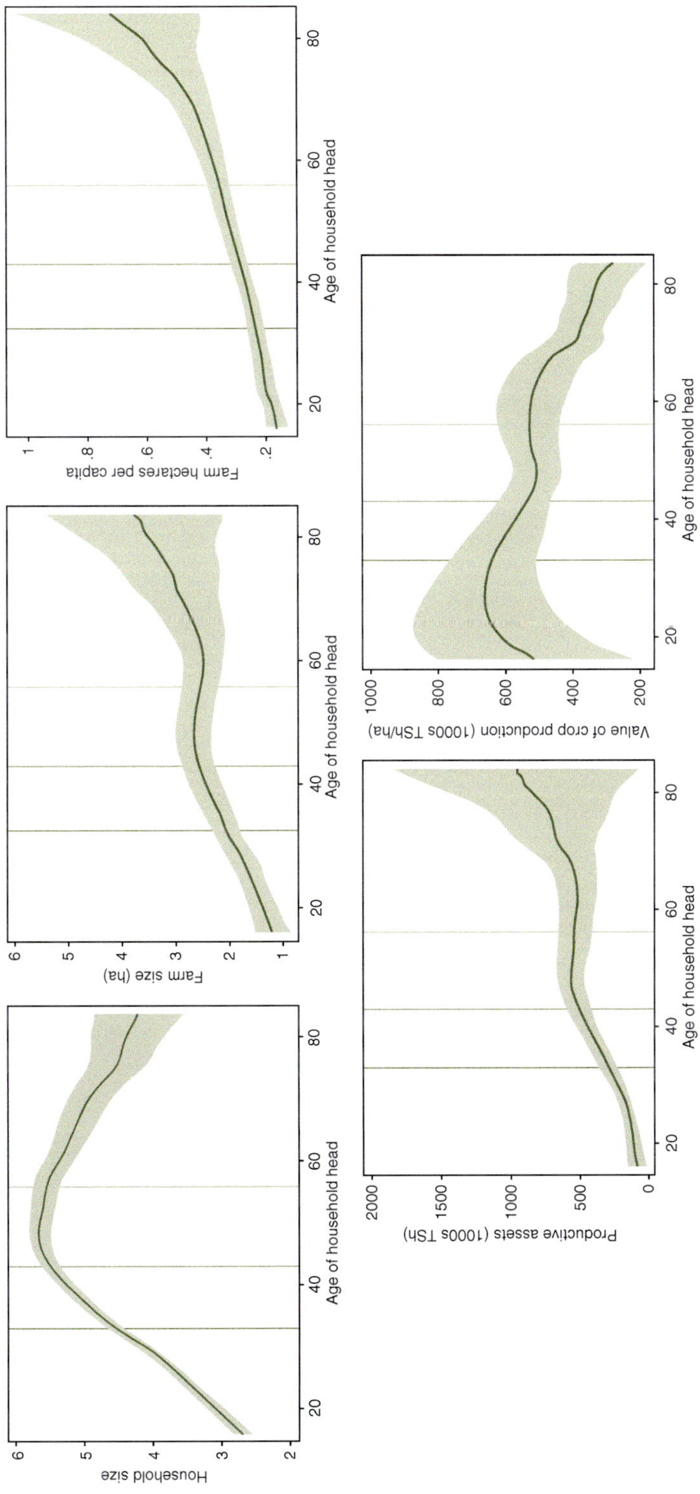

Fig. 6.1a. Selected farm characteristics as affected by age of household head (Tanzania). Vertical lines indicate 25th, 50th and 75th percentiles of household head age distributions. Sample truncated at 99th percentile for graphs. Courtesy of the authors.

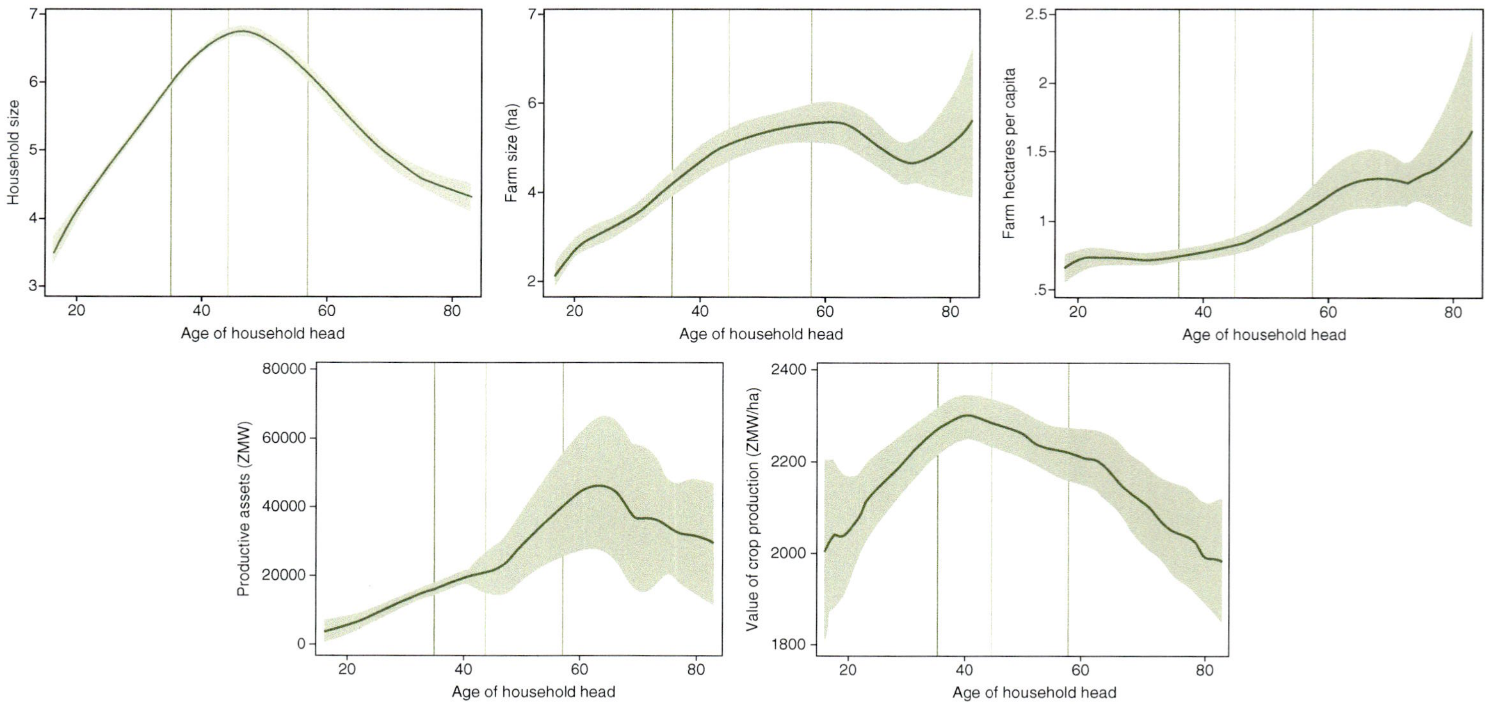

Fig. 6.1b. Selected farm characteristics as affected by age of household head (Zambia). Vertical lines indicate 25th, 50th and 75th percentiles of household head age distributions. Sample truncated at 99th percentile for graphs. Courtesy of the authors.

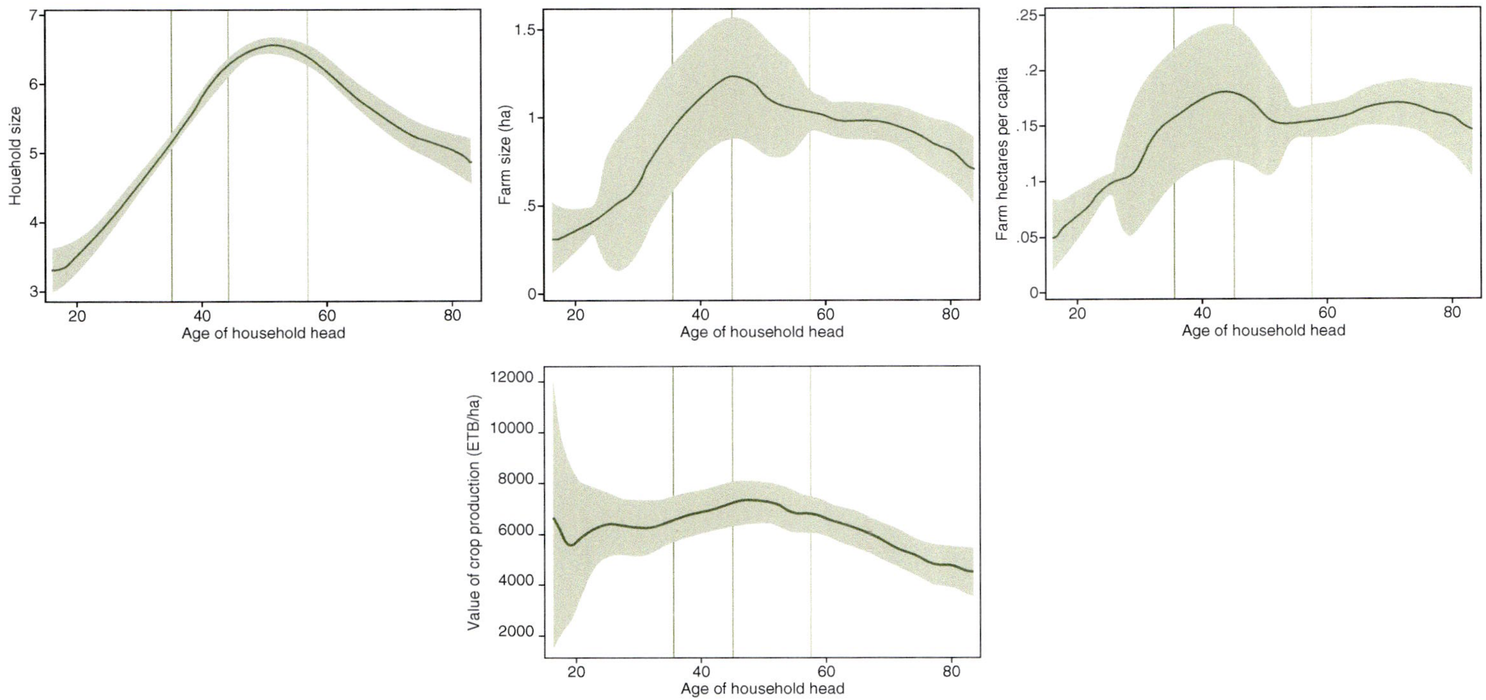

Fig. 6.1c. Selected farm characteristics as affected by age of household head (Ethiopia). Vertical lines indicate 25th, 50th and 75th percentiles of household head age distributions. Sample truncated at 99th percentile for graphs. Courtesy of the authors

Table 6.3a. Descriptive statistics of sample, by quartile of household average adult age, Tanzania.

	Quartile of household average adult age[b]				
Variables	1st	2nd	3rd	4th	Total
Age of household head	31.1	42.6	49.0	61.6	45.3
Avg. age: all members	18.4	21.0	24.7	40.2	25.7
Avg. age: members aged 15+	24.8	30.4	35.6	51.7	35.1
Number of members	4.4	6.1	5.9	4.1	5.1
Number of members aged 15–64	2.7	3.4	3.2	1.9	2.8
Farm size (ha)	2.2	3.5	3.2	2.5	2.8
Number of plots	2.0	2.4	2.4	2.2	2.3
Landless households (%)	48	36	29	20	34
Farm ha/person	0.19	0.32	0.38	0.58	0.36
Value of productive assets[a]	184	350	335	185	259
Value of crop production[a]	285	289	246	173	244
Marketed share of production (%)	36	33	31	23	30
High-value share of production (%)	14	17	17	14	15
Fertilizer (kg/ha)	11.9	16.3	9.5	10.1	11.8
Pesticide (avg. land share) (%)	3	3	3	2	3
Herbicide (avg. land share) (%)	11	11	11	10	10
Uses improved maize seed (%)	56	64	59	52	58
Has irrigation (%)	3	3	5	2	3
Hired labour (%)	44	45	43	43	44

[a]Values in 2015 USD.
[b]Quartiles of sample defined over average age of all household members aged 15 or more.

Table 6.3b. Descriptive statistics of sample, by quartile of household average adult age, Zambia.

	Quartile of household average adult age[b]				
Variables	1st	2nd	3rd	4th	Total
Age of household head	35.2	42.8	48.2	62.6	47.0
Avg. age: all members	17.1	19.4	22.3	37.2	23.9
Avg. age: members aged 15+	25.4	29.8	34.4	49.9	34.7
Number of members	6.0	6.7	6.2	4.4	5.8
Number of members aged 15–64	3.3	3.5	3.1	2.0	3.0
Farm size (ha)	3.9	4.4	4.5	4.5	4.3
Number of plots	3.7	4.0	4.0	3.7	3.8
Landless households (%)	2	1	1	1	2
Farm ha/person	0.69	0.70	0.79	1.29	0.87
Value of productive assets[a]	1831	2073	2827	2944	2413
Value of crop production[a]	238	237	228	211	229
Marketed share of production (%)	38	37	35	30	35
High-value share of production (%)	9	8	8	7	8
Fertilizer (kg/ha)	120.2	117.9	112.6	97.1	112.0
Pesticide (avg. land share) (%)	6.2	5.8	6.1	4.6	5.7
Herbicide (avg. land share) (%)	4.9	5.1	4.7	4.1	4.7
Uses improved maize seed (%)	65	66	64	56	63
Has irrigation (%)	19	20	17	14	17
Mechanized traction (own) (%)	0	0	1	1	1
Mechanized traction (hired) (%)	1	1	1	1	1
Animal traction (own) (%)	23	25	24	20	23
Animal traction (hired) (%)	1	1	1	1	1
Hired labour (%)	38	39	37	39	38

Continued

Table 6.3b. Continued

	Quartile of household average adult age[b]				
Variables	1st	2nd	3rd	4th	Total
Practices conservation tillage (%)	25	27	29	27	27
Practices crop rotation (%)	43	47	46	44	45
Practices intercropping (%)	11	11	11	9	10
Uses crop residues (%)	49	51	51	51	51
Uses mulch (%)	12	14	15	13	14
Has erosion control structures (%)	24	24	23	20	23
Practices agroforestry (%)	5	5	4	4	5

[a] Values in 2015 USD.
[b] Quartiles of sample defined over average age of all household members aged 15 or more.

Table 6.3c. Descriptive statistics of sample, by quartile of household average adult age, Ethiopia.

	Quartile of household average adult age[b]				
Variables	1st	2nd	3rd	4th	Total
Age of household head	33.7	42.6	49.0	60.5	46.4
Avg. age: all members	19.0	21.4	24.7	38.8	25.8
Avg. age: members aged 15+	25.4	30.2	35.1	49.7	34.9
Number of members	5.4	6.3	6.2	4.5	5.7
Number of members aged 15–64	3.3	3.6	3.4	2.1	3.1
Farm size (ha)	0.7	1.4	1.1	0.6	1.0
Number of plots	8.4	9.9	9.8	8.2	9.2
Landless households (%)	37	23	23	26	27
Farm ha/person	0.11	0.21	0.16	0.13	0.15
Value of crop production[a]	397.6	359.4	307.8	267.9	331.3
Marketed share of production (%)	10	9	8	7	8
High-value share of production (%)	20	21	19	17	19
Fertilizer (kg/ha)	443.1	252.0	786.4	145.5	414.3
Pesticide (avg. land share) (%)	8	9	8	6	8
Herbicide (avg. land share) (%)	24	22	23	15	21
Uses improved maize seed (%)	12	13	13	9	12

[a] Values in 2015 USD.
[b] Quartiles of sample defined over average age of all household members aged 15 or more.

year dummies and administrative control dummies (at the regional level for Tanzania and Ethiopia, and at the district level in Zambia).[xi] The full estimation output for all specifications is provided in the Appendix Tables 6A–6C.

Beginning with Tanzania, we find a strong negative correlation between all measures of household age and the practices and commercial outcomes of interest. Comparing specifications using household head versus average age of the family workforce (rows a and b), we find that the coefficient estimates are consistently larger for the household average adult age than for the household head age (although the overall magnitude is small, a topic which we address further below). For example, in the model of improved maize seed usage (column 4), the estimated coefficient for average adult age is twice the size of the coefficient for the age of the household head. This finding is consistent with the idea that some decision making of young household members other than the head is important for the practices on which we have data. When we simultaneously control for age of head

Table 6.4. Influence of household age on farm orientation.

	Variables	(1) Fertilizer (=1)	(2) Pesticide (=1)	(3) Herbicide (=1)	(4) Improved maize seed (=1)	(5) Commercialized share of prod.	(6) High-value share of prod.
	Tanzania						
(a)	Age of household head (years)	–0.000810**	–0.000320**	–0.000908***	–0.00172***	–0.00345***	–0.00103***
		(0.000349)	(0.000129)	(0.000204)	(0.000659)	(0.000286)	(0.000236)
(b)	Avg. age of household adults	–0.00136***	–0.000500***	–0.00127***	–0.00359***	–0.00387***	–0.00139***
		(0.000447)	(0.000165)	(0.000279)	(0.000848)	(0.000349)	(0.000286)
(c)	Age of household head (years)	–0.000396	–0.000267*	–0.000738***	–0.000800	–0.00333***	–0.000813***
		(0.000402)	(0.000159)	(0.000240)	(0.000849)	(0.000335)	(0.000281)
	Avg. age of other household adults	–0.000345	4.94e-05	–0.000545*	–0.00209*	–0.000709*	–0.000621*
		(0.000473)	(0.000222)	(0.000327)	(0.00113)	(0.000405)	(0.000318)
	Zambia						
(a)	Age of household head (years)	–5.35e-05	–0.000438**	–7.75e-05	–0.000472	–0.00150***	–0.000690***
		(0.000319)	(0.000198)	(0.000138)	(0.000369)	(0.000194)	(0.000174)
(b)	Avg. age of household adults	0.000714	–0.000982***	0.000137	–0.000177	–0.00162***	–0.000972***
		(0.000436)	(0.000272)	(0.000179)	(0.000485)	(0.000269)	(0.000241)
(c)	Age of household head (years)	–0.000777*	–3.50e-05	–0.000278	–0.000971**	–0.00138***	–0.000460**
		(0.000398)	(0.000246)	(0.000200)	(0.000443)	(0.000233)	(0.000195)
	Avg. age of other household adults	0.00109**	–0.000697**	0.000548*	0.000560	–0.000252	–0.000144
		(0.000545)	(0.000341)	(0.000297)	(0.000558)	(0.000340)	(0.000278)
	Ethiopia						
(a)	Age of household head (years)	0.000121	–0.000401	–0.00133**	0.000213	–0.000519*	5.16e-05
		(0.000403)	(0.000393)	(0.000564)	(0.000528)	(0.000277)	(0.000266)
(b)	Avg. age of household adults	–0.000509	–0.00111**	–0.00246***	–0.000108	–0.00108***	–9.64e-07
		(0.000606)	(0.000443)	(0.000817)	(0.000779)	(0.000407)	(0.000466)
(c)	Age of household head (years)	0.000276	–9.47e-05	–0.00103*	0.000297	–0.000472	1.97e-05
		(0.000459)	(0.000463)	(0.000609)	(0.000614)	(0.000296)	(0.000279)
	Avg. age of other household adults	0.000461	–0.00127**	–0.00142	–0.000177	–0.000344	0.000529
		(0.000671)	(0.000553)	(0.000987)	(0.000965)	(0.000467)	(0.000641)

Notes: Each country panel shows coefficient estimates from three sets of specifications: (a) controlling for age of household head only; (b) controlling for average age of all adults in household; and (c) controlling for both age of household head and average age of all other household adults. Other controls include household-, farm- and community-level controls, and district and year dummies. Each set of outcomes for a given specification is estimated using a seemingly unrelated regression (SUR) framework, to address correlation in the error terms associated with each outcome. Cluster robust standard errors, shown in parentheses, are robust to heteroskedasticity and within-cluster correlation. Significance levels are denoted as follows: *= p<0.10, **= p<0.05, ***= p<0.01.

and average age of non-head adults (in row c), we find that while the coefficient for head's age is generally larger and more precisely estimated (as evidenced by small standard errors), the age of other adults does contribute further to explaining outcomes. Simply put, while younger heads do seem to have greater propensities to use modern inputs and have a more commercial orientation, at the same time there appears to be additional impacts of a younger workforce on the use of these practices and on the commercial outcome. In other words, for two households identical in all respects, including the age of the head, but differing in the average age of the household labour force, the household with the younger work force is more likely to use modern inputs and engage more intensively with agricultural markets.

Findings for Zambia are remarkably similar to those for Tanzania. The only qualitative exception is the herbicide model results, which indicate that an older workforce is more likely to use herbicide. This result may reflect the labour saving aspects of herbicide being particularly relevant for households with older members. Findings for the models of use of some inputs (pesticide and herbicide) and for commercialized share of production for Ethiopia are also similar to those for Tanzania and Zambia. The insignificant findings for high-value share of production may have to do with a difficult distinction between high-value and staples crops in Ethiopia, where some staples (e.g. *teff*, *ensete*) have high commercial value and are widely marketed, and crops that are grown primarily for market in other countries (e.g. oilseeds), are also widely grown for home consumption in Ethiopia. Similarly, Ethiopia's input acquisition context is somewhat different to other countries in the region: the vast majority of fertilizer and seed acquisitions by smallholders are via a highly regulated and centralized cooperative system, rather than via the private sector. Still, in comparing rows a and b, it is worth noting that not only are the coefficient estimates on age of head uniformly smaller than for the average age of the household workforce, the former are not always negative, while the latter are consistently negative (if not always significant). Here, as in the other countries, it seems that the 'age effect' on smallholder intensification and commercialization is not confined to the named household head.

These results are consistent with younger households being more likely to adopt modern practices and commercial farm orientations than older households, after controlling for resource endowments. However, the age effects we find are exceedingly small in magnitude. Table 6.5 shows estimated values of outcome variables at different ages of the household head (with other variable values at sample means). We find that even the strongest effects are relatively muted. For example, the expected share of marketed crop output for a 20-year-old head is only 7% more than that of a 40-year-old head in Tanzania (and only 3% more in Zambia). Other variables show a much smaller age effect. For example, the likelihood of fertilizer use by a 20-year-old Tanzanian household head is less than 2% greater than that of a 40-year-old head (in Zambia the age difference is not statistically different from zero). Results using average age of adults instead of age of head are of similarly small magnitude.

As farm orientation has multiple dimensions which are not always easy to treat discretely, we attempt to distil some of this variation using factor analysis on the Zambian data. This results in two different composite factors, which we refer to as intensification types A and B: higher scores for intensification type A indicate greater commercialization of staples, fertilizer and improved maize seed usage, while higher scores for intensification type B indicate greater commercialization of high-value crops (like cotton, tobacco and horticultural plants), and pesticide usage. These represent two different modes of intensification–commercialization. Factor loadings for both types are shown in the Appendix Table 6D.

Table 6.6 shows regression results for models in which the scores for factors A and B are the dependent variables. The age of the household head is negatively associated with both, indicating that younger households are more likely to be associated with either orientation.

Another way to examine 'innovativeness' is to examine whether the unexplained portion of variation in farm productivity, after controlling for input use and other observable time-varying conditions affecting production, is associated with farm household age. To do this, we estimated a production function for aggregate value of production per hectare using a fixed effects estimator. The individual fixed effects estimate from this model captures the portion of the variability in productivity outcomes which is associated with the individual, but not otherwise explained

Table 6.5. Predicted farm orientation outcomes for Tanzania and Zambia.

	Variables	(1) Commercialized share of prod.	(2) High-value share of prod.	(3) Fertilizer (=1)	(4) Pesticide (=1)	(5) Herbicide (=1)	(6) Improved maize seed (=1)
	Tanzania						
(a)	Predicted outcome I head age = 20	0.400	0.171	0.180	0.041	0.128	0.491
(b)	Predicted outcome I head age = 30	0.366	0.161	0.172	0.038	0.118	0.474
(c)	Predicted outcome I head age = 40	0.331	0.151	0.163	0.034	0.109	0.457
	Difference (b) – (a)	–0.034	–0.010	–0.009	–0.003	–0.009	–0.017
	Difference p-value	(0.000)	(0.000)	(0.016)	(0.013)	(0.000)	(0.010)
	Difference (c) – (a)	–0.068	–0.020	–0.017	–0.006	–0.018	–0.034
	Difference p-value	(0.000)	(0.000)	(0.016)	(0.013)	(0.000)	(0.010)
	Zambia						
(a)	Predicted outcome I head age = 20	0.348	0.195	0.524	0.131	0.0559	0.599
(b)	Predicted outcome I head age = 30	0.332	0.188	0.524	0.128	0.0551	0.591
(c)	Predicted outcome I head age = 40	0.317	0.181	0.523	0.124	0.0544	0.582
	Difference (b) – (a)	–0.015	–0.007	–0.001	–0.004	–0.001	–0.008
	Difference p-value	(0.000)	(0.000)	(0.867)	(0.071)	(0.591)	(0.031)
	Difference (c) – (a)	–0.031	–0.014	–0.001	–0.007	–0.001	–0.017
	Difference p-value	(0.000)	(0.000)	(0.867)	(0.071)	(0.591)	(0.031)

Notes: Each country panel shows predicted outcomes at different ages of household head, and the significance of differences between these predicted outcomes, after controlling for household-, farm- and community-level controls, and district and year dummies. P-values are shown in parentheses.

by model covariates (i.e. observed inputs and management decisions). This fixed effects estimate has sometimes been interpreted as a latent measure of individual farming ability (e.g. Deininger *et al.*, 2013; Jin and Jayne, 2013; Chamberlin and Ricker-Gilbert, 2016). We then calculated the correlation between this 'ability' proxy and our various measures of household age (Table 6.7). The correlation coefficients are all negative and highly significant, indicating that unobserved factors associated with land productivity are negatively associated with household age. One interpretation of this is that younger households are more productive, although in ways which are difficult to measure directly through existing household survey data.

Much effort and resources have been expended to promote the use of climate-smart approaches and sustainable intensification (SI) production techniques in Zambia in recent years, although it should be noted that some of these same practices have been promoted over decades. To the extent that these practices represent non-traditional management choices, they may be taken as measures of 'doing things differently'. Table 6.8 shows regression results for models in which the dependent variable is a farm-level measure of use of: (1) minimum or zero tillage; (2) grain–legume crop rotation; (3) intercropping; (4) crop residue retention; (5) mulching; (6) erosion control structures, such as bunds or terraces; and (7) agroforestry. All practices were measured jointly in a seemingly unrelated regressions (SUR) framework to allow for correlated error terms. Results indicate that age of head is positively associated with the use of most of these practices – agroforestry is the single exception. This suggests that younger households are not generally more likely to engage in such practices. However, it should be acknowledged that if SI

Table 6.6. Correlates of factor scores (farm orientation) for Zambian data.

Variables	Intensification type A	Intensification type B
Age of household head (years)	−0.00299***	−0.00368***
	(3.34e-06)	(1.33e-09)
Max. educational attainment	−0.0129***	0.0744***
	(0.000132)	(0)
Female head (1/0)	−0.132***	−0.0815***
	(3.90e-08)	(0.000942)
Farm size (ha)	0.00406***	−0.000201
	(0.00135)	(0.901)
Household members (count)	0.0181**	−0.0122
	(0.0293)	(0.147)
Productive assets (ZMW)	2.64e-07**	−1.31e-07
	(0.0269)	(0.249)
Hours to town of 50k+	3.66e-05	−6.93e-05***
	(0.181)	(0.00609)
Population density	0.382***	0.580***
	(0.000302)	(0)
Rainfall (mm)	0.000280***	0.000510***
	(0.00380)	(4.02e-09)
Intraseasonal rainfall CV	0.798***	−0.0566
	(3.21e-05)	(0.708)
Observations	16,122	16,122
R-squared	0.272	0.312
District FE	Yes	Yes
Year FE	Yes	Yes
Mundlak–Chamberlain device	Yes	Yes

Notes: Cluster robust standard errors, shown in parentheses, are robust to heteroskedasticity and within-cluster correlation. Significance levels are denoted as follows: **= p<0.05, ***= p<0.01. FE = fixed effects

Table 6.7. Correlation between unobserved household farming 'ability' and age, for Zambian data.

Variables	Correlation coefficient
Age of household head	−0.0417***
Average adult age	−0.0649***
Average adult age (excluding head)	−0.0307***

Notes: Adults are members aged ≥15. 'Ability' is the recovered fixed effect estimate from a production function using the fixed effects estimator. Significance levels are denoted as follows: ***= p<0.01.

promotion preferentially targets older farmers, then we might expect such associations, even if younger households might be more receptive to new practices in general.

Finally, much has been said about young people's affinity for modern technologies, including mobile phones and other information and communications technologies (ICTs) (FAO *et al.*, 2014; AGRA, 2015). In our data set, we observe mobile phone usage reported along with the main purpose of usage. Table 6.9 shows regression results for binary indicators of whether or not a household (1) uses a mobile phone for any purpose; (2) uses a mobile phone to call family and friends; or (3) uses a mobile phone for business activities, such as sending or receiving money, negotiating sales transactions or obtaining market information. In the model, the dependent variable in column 4 is the number of business activities for which a mobile phone is used. Coefficient estimates indicate that households with younger heads are more likely to report using mobile phones than those with older heads; they are more likely to use mobile phones for business activities; and on average younger household heads use mobile phones for a greater variety of business activities. These results could be read as

Table 6.8. Correlates of sustainable land management practices in Zambia.

	(1)	(2)	(3)	(4)	(5)	(6)	(7)
Variables	Min./zero tillage	Crop rotation	Intercropping	Crop residue	Mulching	Erosion control	Agroforestry
Age of household head (years)	0.00107***	0.00141***	0.000222	0.00126***	0.000504*	0.000197	−0.000515***
	(0.000332)	(0.000300)	(0.000282)	(0.000423)	(0.000286)	(0.000336)	(0.000182)
Max. educational attainment	0.00282*	0.00517***	0.00230*	0.00346	0.00217*	0.00166	0.00144
	(0.00163)	(0.00163)	(0.00131)	(0.00230)	(0.00128)	(0.00159)	(0.00104)
Female head (1/0)	−0.0169	0.0132	0.0128	−0.00535	−0.0126	−0.0242*	0.00352
	(0.0136)	(0.0120)	(0.00959)	(0.0166)	(0.00942)	(0.0131)	(0.00709)
Farm size (ha)	−0.000129	0.00175*	−0.000103	−0.000571	0.000181	−1.60e-05	−0.00108**
	(0.000364)	(0.00103)	(0.000316)	(0.000445)	(0.000325)	(0.000280)	(0.000471)
Household members (count)	−0.00616**	−0.00125	0.00222	0.00508	0.00133	0.00149	−0.00298
	(0.00298)	(0.00522)	(0.00172)	(0.00329)	(0.00155)	(0.00222)	(0.00435)
Productive assets (ZMW)	6.02e-09	6.25e-08	−1.77e-08	−4.33e-08	1.54e-08	−1.56e-08	2.92e-08
	(1.46e-08)	(5.69e-08)	(1.35e-08)	(2.95e-08)	(1.28e-08)	(1.96e-08)	(3.97e-08)
Hours to town of 50k+	2.31e-06	4.70e-06	6.15e-06	−1.88e-05	3.76e-06	−1.10e-05	2.36e-05***
	(1.10e-05)	(1.34e-05)	(9.45e-06)	(1.40e-05)	(9.04e-06)	(1.01e-05)	(7.16e-06)
Population density	0.00450	0.0547*	0.0135	0.123***	0.0850***	0.0673***	0.0399***
	(0.0458)	(0.0286)	(0.0326)	(0.0255)	(0.0235)	(0.0223)	(0.00944)
Rainfall (mm)	−0.000110***	7.30e-05	3.71e-05*	5.53e-07	−6.61e-05*	1.20e-05	−0.000126*
	(3.35e-05)	(7.35e-05)	(2.22e-05)	(5.40e-05)	(3.56e-05)	(2.44e-05)	(7.31e-05)
Intraseasonal rainfall CV	−0.0322	−0.173	0.190***	0.179**	0.147***	0.0196	−0.0896
	(0.0724)	(0.118)	(0.0480)	(0.0822)	(0.0541)	(0.0450)	(0.150)
Constant	0.263	−0.425	−0.217	−0.266	−0.811***	−0.521**	−0.454***
	(0.379)	(0.259)	(0.290)	(0.292)	(0.206)	(0.257)	(0.116)
Observations	16,454	16,454	16,454	16,454	16,454	16,454	16,454
Mundlak–Chamberlain device	Yes	Yes	Yes	Yes	Yes	Yes	Yes
District FE	Yes	Yes	Yes	Yes	Yes	Yes	Yes
Year FE	Yes	Yes	Yes	Yes	Yes	Yes	Yes

Notes: The average age of other household adults is the average age of all members, other than the head, who are aged ≥15. Cluster robust standard errors, shown in parentheses, are robust to heteroskedasticity and within-cluster correlation. Significance levels are denoted as follows: *= p<0.10, **= p<0.05, ***= p<0.01. FE = fixed effects.

Table 6.9. Correlates of mobile phone usage in Zambia.

	(1)	(2)	(3)	(4)
Variables	1=use mobile phone for any purpose	1=use mobile phone to call family or friends	1=use mobile phone for any business activity	# of business activities for which mobile phone used
Age of household head (years)	–0.00152***	0.000624	–0.00234***	–0.00804***
	(0.000543)	(0.000385)	(0.000529)	(0.00138)
Max. educational attainment	0.0245***	–0.00205	0.0296***	0.109***
	(0.00251)	(0.00177)	(0.00234)	(0.00739)
Female head (1/0)	–0.0421**	0.0277**	–0.0746***	–0.199***
	(0.0197)	(0.0127)	(0.0190)	(0.0494)
Farm size (ha)	–0.00288*	–0.000235	–0.00247	–0.0196***
	(0.00156)	(0.000869)	(0.00158)	(0.00649)
Household members (count)	–0.0247***	–0.00811	–0.0179**	–0.0151
	(0.00898)	(0.00630)	(0.00857)	(0.0254)
Productive assets (ZMW)	1.96e-07	7.79e-08	1.64e-07	9.53e-07*
	(1.61e-07)	(4.95e-08)	(1.51e-07)	(5.57e-07)
Hours to town of 50k+	–7.10e-05***	2.71e-05*	–0.000106***	–0.000288***
	(2.14e-05)	(1.51e-05)	(2.08e-05)	(6.03e-05)
Population density	–0.0845*	–0.0215	–0.0943**	–0.0445
	(0.0434)	(0.0321)	(0.0399)	(0.145)
Rainfall (mm)	0.000472	7.84e-05	0.000577**	0.00150**
	(0.000305)	(0.000218)	(0.000286)	(0.000726)
Intraseasonal rainfall CV	–0.238	0.0440	–0.235	0.159
	(0.555)	(0.379)	(0.508)	(1.569)
Observations	6813	6813	6813	7747
R-squared	0.164	0.215	0.140	0.172
Mundlak–Chamberlain device	Yes	Yes	Yes	Yes
District FE	Yes	Yes	Yes	Yes

Notes: The average age of other household adults is the average age of all members, other than the head, who are aged ≥15. Cluster robust standard errors, shown in parentheses, are robust to heteroskedasticity and within-cluster correlation. Significance levels are denoted as follows: *= p<0.10, **= p<0.05, ***= p<0.01. FE = fixed effects.

lending themselves to the conventional narrative of young farmers being more engaged with ICT in rural Africa, but caution is advised as we know little about who within the household is actually using the phone.

Qualitative research undertaken as part of the International Fund for Agricultural Development (IFAD) Youth Research project across four sites in each of four SSA countries (see Chapter 2, this volume) supports the conclusion that there is little evidence for an ongoing, youth-led rural economic revolution. The qualitative research does not provide much support for the idea that young people are using different agricultural technology or using technology differently. Further, while many young people market some of their agricultural produce, there is little evidence that young people are engaged in innovative processing activities or linking to value chains characterized by higher value products, value addition, safety requirements and quality differentiation (Reardon, 2015).

More broadly, while most young women and young men engage in one or more non-farm income-generating activities, overwhelmingly these tend to be the same low-investment, low-technology, and relatively low-skill activities, including petty trading, food preparation, catering and artisanal trades, which dominate the rural opportunity landscape (see also Flynn and Sumberg, 2018; Yeboah *et al.*, 2020; Chapter 3, this volume). One suggestion is that even in areas of significant agricultural commercialization, there are still relatively few meaningful opportunities for investment, skill upgrading and risk taking. Another is that the intergenerational nature of much farm and non-farm economic activity, and the importance of social relations for accessing land, capital and skill, effectively blur the lines between youth and non-youth economic activity.

Discussion

Our empirical evaluation in the preceding section suggests two conclusions which stand somewhat in counterpoint to each another. On the one hand, we do find fairly pronounced and widespread evidence that younger households are more likely to adopt modern practices and have more commercial orientations, compared with older households, and despite the observable difference in resource endowments that also characterize young rural households. However, these effects are generally quite small in magnitude, even when the estimates are fairly precise. This suggests that rhetoric about unleashing a whirlwind of youth-led agricultural transformation is probably misplaced. Education, specialized training and access to credit and other investment resources may play important roles in facilitating young farmers to realize their productive potential – but the same could certainly also be said for older adults.

Our finding that household average adult age seems to offer more explanatory power than the age of the household head suggests that the default position of measuring 'farmer characteristics' in terms of the nominal household head requires further critical scrutiny. We feel it is very likely that distributed models of decision making may better map on to the reality of smallholder households and their various agricultural and non-farm activities. This is perhaps particularly true of contemporary SSA, where land constraints may keep adult children within natal households for longer than they might desire, and where traditional cultural norms may allocate nominal headship to individuals based on seniority, gender and genealogical position, even when the farm management aspects of headship have devolved to other members. There is some qualitative evidence that suggests this is the case, e.g. Andersson Djurfeldt *et al.*'s (2019) study on household positionality and intrahousehold participation in farm management in Zambia. Our argument in this respect is similar to arguments for more nuanced and creative collection of data on gendered patterns of intrahousehold resource control and decision making (Doss *et al.*, 2018) and more general critiques of uncritical adoption of unitary household models of decision making (Agarwal, 1997). The intrahousehold resource allocation literature has focused much of its attention on gendered access to and control over household assets and the implications for individual and household welfare outcomes (e.g. Doss, 2013; Oduro *et al.*, 2015). Similar critical reinterpretations of intrahousehold agency and dynamics with respect to farm management and production – and the role that young adults play therein – may be very illuminating. This will require more creative approaches to data collection. Recent efforts to collect information from multiple contributors to decision making are a step in the right direction, but still offer limited insights into the agency of non-head members, and may poorly reflect more complex situations of collective intrahousehold agency and collaborative managerial decisions. Another issue is that cultivation plots which are entirely controlled by subordinate members of a household may not even show up in household rosters. Qualitative evidence such as that provided by Andersson Djurfeldt *et al.* (2019) suggests this is not unlikely for some types of households.[xii] If we are serious about understanding how young people farm (and do other things), it is incumbent on us to do more critical evaluation of how well standard household instruments capture individual-level and collective economic activity.

We should acknowledge that the regression models we employ in the analysis for this chapter use continuous linear measures of age, rather than discrete categories of 'young'. As such, the modelling portion of our analysis may be characterized as the exploration of age effects, rather than youth effects per se, in conditioning our primary outcomes of interest. We may note, however, that we did estimate alternative specifications for these models with discrete 'youth' dummies and the overall story is very consistent with the results presented here. As the discrete definition of a 'youth' indicator is fundamentally arbitrary (is 15–25 years of age a better definition than 15–28? Should we use the same definition of youth for heads and for household labour?), we have elected to restrict our discussion in this chapter to the linear models as presented. We also estimated models with non-linear measures of age, but generally found that linear models performed as well or better and are more easily interpreted, particularly in the specification which includes the age of the head and the average age of the household adults. Further empirical work may examine alternative analytical approaches in more detail.

Another important point relates to the nature of the farming activities and outcomes we are able to examine with survey data. Ideally, we would like to say something about innovation, entrepreneurial disruption and the like, but we are limited in our ability to observe these. It may be the case that youth-driven changes are taking place that we simply are unable to observe because we are not asking the right questions. While this is a possibility, it would be a mistake to simply assume it is the case.

Finally, it is important to recognize that we only observe those who stay in farming. As such, our results are representative only of those who are actually farming, not of all the young people who might have farmed if things had been different (e.g. more access to land, fewer competing opportunities in urban areas). The individuals with the most transformative potential may not be the ones starting farm households (or staying on their parents' farms). That said, the standard conceptualization of the structural transformation process suggests that the least efficient farmers will be the first to leave (Johnston and Mellor, 1961). If the latter is the case, then policies that reduce barriers to farming (e.g. via land market promotion, resettlement schemes, credit facilitation) may induce entrance by individuals with lower inherent productive or innovative capacity. In any case, observational data may never be able to fully address the selection bias at work here, and there may be limited scope for experimental approaches. Nevertheless, thinking about this question more carefully may clarify new opportunities for empirical research.

Conclusions

We find some support for many of the stylized assertions about youth in African agriculture. Younger household heads have smaller farms, smaller households and fewer capital resources. However, after controlling for these constraints, we observe marginally higher propensities to engage with intensification practices and commercial orientations which are objectives of much current policy and programming. These findings are in line with the narrative of young people's inherent vim and vigour being an asset for agricultural growth and transformation in SSA.

However, the very limited magnitude of these age effects suggests that much caution should be exercised in making this argument. More work is needed to understand the conditions under which young farmers thrive. It is not clear that policies or programmes that target the land, capital or other endowments and constraints of young farmers, will have bigger impacts than transformative investments that are accessible by all farmers. It is probably not useful (and certainly not realistic) to divorce young people from their resource constraints: the implications of this are that there are probably few viable policy avenues for unleashing the much vaunted, latent youth whirlwind of innovative effervescence.

In any case, the locus of change in rural economies may be happening off the farm, and in ways that are opaque to our current ways of monitoring the system (i.e. household surveys). Examples include medium-scale farm investment; agribusiness and value chain expansion; urban demand and the supermarket revolution; and lower barriers to global agricultural markets. It is not at all clear what role youth may play (if any) in such areas, but new ways and a different focus of research and data collection may allow that question to be probed to a further extent than we are able to here.

Notes

[i] www.mastercardfdn.org/yaw2017-infographic/ (accessed 26 November 2020).

[ii] To be fair, most assessments of rural youth out-migration evaluate it as an essentially positive part of the structural transformation process, although the potential for negative impacts on agricultural productivity is mentioned in many reports, usually operating through loss of productive labour (e.g. The World Bank, 2015; Deotti and Estruch, 2016; IADB *et al.*, 2017; FAO *et al.*, 2018).

[iii] www.gemconsortium.org/ (accessed 26 November 2020).

[iv] See Margolis (2014) for discussion of some of the data-related reasons underlying such conflation.

[v] By 'commercial orientation', we refer to the degree to which the farm production portfolio is organized around generating marketable surplus (as opposed to satisfying household consumption requirements). Unless otherwise specified, we use a common shorthand measure of this: the share of total value of crop production which is sold (sometimes referred to as the 'household commercialization index').

[vi] Attitudes toward innovation, and the propensity to innovate, are ideas that partially overlap with other personality features, such as risk aversion, creativity and receptivity to new ideas. A number of empirical measures of risk aversion – a trait that maps on to some notions of entrepreneurial disposition – have been developed. These are not part of most large-scale household survey data sets, however, and so we do not discuss them in detail. Creativity and openness to new ideas are similarly measurable in principle, although typically not part of available survey data.

[vii] See Hammond *et al.* (2017) for discussion of measurement challenges with tracking climate-smart agricultural practices with existing household survey instruments.

[viii] Data are from the 2015 Zambian Rural Agricultural Livelihoods Survey (RALS) (CSO, 2012) RALS, 2012–2013 Tanzanian LSMS–ISA, and 2015–2016 Ethiopian LSMS–ISA surveys. For Zambia, responses to the question, 'Who primarily decided how to use this field?' were asked for all fields reported by the household. For Tanzania, we report the first response to the question, 'Who decided what to plant on this plot in the long rainy season?' asked for cropped plots (up to three members were possible joint decision makers). For Ethiopia, we report the response to the question, 'Who in the household makes primary decisions concerning crops to be planted, input use and the timing of cropping activities on this field?' asked for cropped plots.

[ix] The LSMS–ISA data were produced as part of a World Bank project (www.worldbank.org/en/programs/lsms/initiatives/lsms-ISA, accessed 26 November 2020), in collaboration with the Ethiopian Central Statistics Agency (CSA) and the Tanzanian National Bureau of Statistics (NBS) in each of those respective countries. Detailed descriptions of instrument design, sampling frame and other aspects of data collection are provided for Ethiopia by CSA (2017) and for Tanzania by NBS (2014). The Zambian RALS was produced by the Indaba Agricultural Policy Research Institute (IAPRI), collaborating with the Zambia Central Statistical Office (CSO) and the Ministry of Agriculture (MoA). For more details on RALS questionnaire and sampling design, see CSO (2012).

[x] High-value crops were defined as all non-staple crops, i.e. horticultural crops, high-value oilseeds, food crops such as sugarcane, and non-food crops such as tobacco and cotton.

[xi] We also include time averages of time-varying household characteristics to control for unobserved time-varying heterogeneity, which may bias our results. This formulation, sometimes referred to as the 'Mundlak–Chamberlain device', is based on the assumption that time averages are correlated with unobserved time-invariant factors (Mundlak, 1978; Chamberlain, 1984; Wooldridge, 2010). This estimator is referred to as the 'correlated random effects model', showing that it provides consistent estimates under the conditionality of its assumptions. Each set of outcomes for a given specification is estimated using a seemingly unrelated regression (SUR) framework, to address correlation in the error terms associated with each outcome. Standard errors are robust to clustering at the enumeration level.

[xii] '[One] variety of positionality involves young men and their wives who live in extended, multi-generational families, where they together with their children constitute the middle segment of these families. They have been allocated family land by their parents and farm independently but continue to live close to the parental generation and view the patriarch as the household head, even if he is not involved in any of their production or marketing decisions and they are not accountable to him in terms of incomes raised and saved' (Andersson Djurfeldt *et al.*, 2019 p.7), and 'a handful of respondents were also single, unmarried men who had been allocated land by their parents and continued living with their parents, but were farming independently and were not accountable to their parents' (Andersson Djurfeldt *et al.*, 2019). Given the low incidence of plots identified as controlled by members other than head and/or spouse in household survey data, it would seem that the types of households referred to above may not fully report on the land managed by all individuals identified as household members.

References

Africa Commission 2009. *Realising the potential of Africa's youth: Report of the Africa Commission*, Copenhagen, Ministry of Foreign Affairs of Denmark.

Agarwal, B. 1997. 'Bargaining' and gender relations: Within and beyond the household. *Feminist Economics*, 3, 1–51.

AGRA 2015. *Africa Agriculture Status Report 2015: Youth in Agriculture in Sub-Saharan Africa*, Nairobi, Alliance for Green Revolution in Africa (AGRA).

Allan, W. 1965. *The African Husbandman*, London, Oliver & Boyd.

Andersson Djurfeldt, A., Kalindi, A., Lindsjö, K. & Wamulume, M. 2019. Yearning to farm – Youth, agricultural intensification and land in Mkushi, Zambia. *Journal of Rural Studies*, 71, 85–93.

Biggs, S. 1990. A multiple source of innovation model of agricultural research and technology promotion. *World Development*, 18, 1481–1499.

Chamberlain, G. 1984. Panel data. *In*: Grilliches, Z. & Intriligator, M. D. (eds.) *Handbook of Econometrics*, Volume 2. Amsterdam, North-Holland Press.

Chamberlin, J. & Ricker-Gilbert, J. 2016. Participation in rural land rental markets in sub-Saharan Africa: who benefits and by how much? Evidence from Malawi and Zambia. *American Journal of Agricultural Economics*, 98, 1507–1528.

CSA 2017. *LSMS–Integrated Surveys on Agriculture Ethiopia Socioeconomic Survey (ESS) 2015/2016: A Report by the Central Statistical Agency of Ethiopia in Collaboration with the National Bank of Ethiopia and The World Bank*, Addis Ababa, Ethiopia, Central Statistical Agency.

CSO 2012. *The 2012 Rural Agricultural Livelihoods Survey (RALS) Interviewer's Instruction Manual. Central Statistical Office (Agriculture and Environment Division) in conjunction with the Ministry of Agriculture and Livestock (Policy and Planning Department), commissioned by Michigan State University (MSU) and Indaba Agricultural Policy Research Institute (IAPRI)*, Lusaka, Zambia, Central Statistical Office.

Deininger, K., Ali, D. A. & Alemu, T. 2013. Productivity effects of land rental market operation in Ethiopia: evidence from a matched tenant–landlord sample. *Applied Economics*, 45(25), 3531–3551.

Deotti, L. & Estruch, E. 2016. *Addressing Rural Youth Migration at its Root Causes: A Conceptual Framework*, Rome, Food and Agriculture Organisation (FAO).

Diao, X., Fang, P., Magalhaes, E., Pahl, S. & Silver, J. 2019. Cities and rural transformation: A spatial analysis of rural youth livelihoods in Ghana. *In*: Mueller, V. & Thurlow, J. (eds.) *Youth and Jobs in Rural Africa: Beyond Stylized Facts*, Oxford, UK, Oxford University Press.

Doss, C. 2013. Intrahousehold bargaining and resource allocation in developing countries. *The World Bank Research Observer*, 28, 52–78.

Doss, C., Oduro, A. D., Deere, C. D., Swaminathan, H., Baah-Boateng, W. & Suchitra. J.Y. 2018. Assets and shocks: a gendered analysis of Ecuador, Ghana and Karnataka, India. *Canadian Journal of Development Studies / Revue Canadienne d'Etudes du Développement*, 39, 1–18.

FAO, CTA & IFAD 2014. *Youth and Agriculture: Key Challenges and Concrete Solutions*, Rome, Food and Agriculture Organization of the United Nations (FAO) in collaboration with the Technical Centre for Agricultural and Rural Cooperation (CTA) and the International Fund for Agricultural Development (IFAD).

FAO, IFAD, IOM & WFP 2018. *The Linkages between Migration, Agriculture, Food Security and Rural Development*, Rome, Food and Agriculture Organization (FAO), International Fund for Agricultural Development (IFAD), International Organization for Migration (IOM) and World Food Programme (WFP).

Flynn, J. & Sumberg, J. 2018. Are savings groups a livelihoods game changer for young people in Africa? *Development in Practice*, 28, 51–64.

Glover, D. & Sumberg, J. 2020. Youth and food systems transformation. *Frontiers in Sustainable Food Systems*, 4 (doi.org/10.3389/fsufs.2020.00101).

Glover, D., Sumberg, J., Ton, G., Andersson, J. & Badstue, L. 2019. Rethinking technological change in smallholder agriculture. *Outlook on Agriculture*, 48, 169–180.

Gough, K. V. & Langevang, T. 2017. Introduction: Youth entrepreneurship in sub-Saharan Africa. *In*: Gough, K. V. & Langevang, T. (eds.) *Young Entrepreneurs in Sub-Saharan Africa*, London, Routledge.

Guo, Q., Ola, O. & Benjamin, E. O. 2020. Determinants of the adoption of sustainable intensification in southern African farming systems: A meta-analysis. *Sustainability*, 12, Paper 3276.

Hammond, J., Fraval, S., Van Etten, J., Suchini, J. G., Mercado, L., Pagella, T., Frelat, R., Lannerstad, M., Douxchamps, S., Teufel, N., Valbuena, D. & Van Wijk, M. T. 2017. The Rural Household Multi-Indicator Survey (RHoMIS) for rapid characterisation of households to inform climate smart agriculture interventions: Description and applications in East Africa and Central America. *Agricultural Systems*, 151, 225–233.

Hill, P. 1963. *Migrant Cocoa-Farmers of Southern Ghana*, Cambridge, UK, Cambridge University Press.

IADB, IFAD, IOM, OAS & WFP 2017. *Food Security and Emigration. Why People Flee and the Impact on Family Members Left Behind in El Salvador, Guatemala and Honduras*, Rome, World Food Programme (WFP).

Jin, S. & Jayne, T. S. 2013. Land rental markets in Kenya: implications for efficiency, equity, household income, and poverty. *Land Economics*, 89, 246–271.

Johnston, B. F. & Mellor, J. W. 1961. The role of agriculture in economic development. *The American Economic Review*, 51, 566–593.

Kassie, M., Teklewold, H., Jaleta, M., Marenya, P. & Erenstein, O. 2015. Understanding the adoption of a portfolio of sustainable intensification practices in eastern and southern Africa. *Land Use Policy*, 42, 400–411.

Kew, J., Namatovu, R., Aderinto, R. & Chigunta, F. 2015. *Africa's Young Entrepreneurs: Unlocking the Potential for a Brighter Future*, Rondebosch, South Africa, Development Unit for New Enterprise, University of Cape Town.

Knowler, D. & Bradshaw, B. 2007. Farmers' adoption of conservation agriculture: A review and synthesis of recent research. *Food Policy*, 32, 25–48.

Krishna, V. V., Yigezu, Y. A., Karimov, A. A. & Erenstein, O. 2020. Assessing technological change in agri-food systems of the Global South: A review of adoption-impact studies in wheat. *Outlook on Agriculture*, 49, 89–98.

Mabiso, A. & Benfica, R. 2019. The narrative on rural youth and economic opportunities in Africa: facts, myths and gaps. *IFAD Research Series* 61 Rome, IFAD.

Margolis, N. D. 2014. By choice and by necessity: Entrepreneurship and self-employment in the developing world. *The European Journal of Development Research*, 26, 419–436.

Meijer, S. S., Catacutan, D., Ajayi, O. C., Sileshi, G. W. & Nieuwenhuis, M. 2015. The role of knowledge, attitudes and perceptions in the uptake of agricultural and agroforestry innovations among smallholder farmers in sub-Saharan Africa. *International Journal of Agricultural Sustainability*, 13, 40–54.

Mundlak, Y. 1978. On the pooling of time series and cross section data. *Econometrica*, 46, 69–85.

NBS 2014. *Tanzania National Panel Survey Report (NPS) – Wave 3, 2012–2013*, Dar es Salaam, Tanzania, National Bureau of Statistics (NBS).

Ngoma, H., Pelletier, J., Mulenga, B. P. & Subakanya, M. 2020. *Climate-Smart Agriculture, Cropland Expansion, and Deforestation in Zambia: Linkages, Processes, and Drivers (No. 303673)*, East Lansing, MI, Michigan State University, Department of Agricultural, Food, and Resource Economics, Feed the Future Innovation Lab for Food Security (FSP).

Niosi, J., Saviotti, P., Bellon, B. & Crow, M. 1993. National systems of innovation: in search of a workable concept. *Technology in Society*, 15, 207–227.

Oduro, A. D., Deere, C. D. & Catanzarite, Z. B. 2015. Women's wealth and intimate partner violence: Insights from Ecuador and Ghana. *Feminist Economics*, 21, 1–29.

Okali, C. & Sumberg, J. 2012. Quick money and power: Tomatoes and livelihood building in rural Brong Ahafo, Ghana. *IDS Bulletin*, 43, 44–57.

Reardon, T. 2015. The hidden middle: the quiet revolution in the midstream of agrifood value chains in developing countries. *Oxford Review of Economic Policy*, 31, 45–63.

Richards, P. 1986. *Coping With Hunger: Hazard and Experiment in an African Rice Farming System*, London, Allen & Unwin.

Rogers, E. M. 1983. *The Diffusion of Innovations*, New York, The Free Press.

Schumpeter, J. A. 1947. The creative response in economic history. *Journal of Economic History*, November, 149–159.

Sumberg, J. & Hunt, S. 2019. Are African rural youth innovative? Claims, evidence and implications. *Journal of Rural Studies*, 69, 130–136.

Sumberg, J. & Okali, C. 1997. *Farmers' Experiments: Creating Local Knowledge*, Boulder, CO, Lynne Rienner.

Sumberg, J. & Okali, C. 2006. Tomatoes, decentralization, and environmental management in Brong Ahafo, Ghana. *Society & Natural Resources*, 19, 19–31.

The World Bank 2015. *Women in Agriculture: The Impact of Male Out-migration on Women's Agency, Household Welfare, and Agricultural Productivity. Report No. AUS9147*, Washington, DC, The World Bank.

Wooldridge, J. M. 2010. *Econometric Analysis of Cross Section and Panel Data (2nd ed.)*, Cambridge, MA, MIT Press.

Yeboah, T., Chigumira, E., John, I., Anyidoho, N. A., Manyong, V., Flynn, J. & Sumberg, J. 2020. Hard work and hazard: Young people and agricultural commercialisation in Africa. *Journal of Rural Studies*, 76, 142–151.

Appendix

Table 6A1. Full regression estimates for Tanzania (specification 1, see Table 6.4).

Variables	(1) Commercialized share of prod.	(2) High-value share of prod.	(3) Fertilizer (kg/ha)	(4) Pesticide (=1)	(5) Herbicide (=1)	(6) Improved maize seed (=1)
Age of household head (years)	−0.00344***	−0.00103***	−3.120*	−0.000320**	−0.000908***	−0.00172***
	(0.000286)	(0.000236)	(1.638)	(0.000129)	(0.000204)	(0.000659)
Female head (1/0)	0.0145	−0.0317*	12.22	0.00209	0.0118	−0.00250
	(0.0331)	(0.0184)	(17.49)	(0.0111)	(0.0295)	(0.0767)
Farm size (ha)	0.00393**	0.00198*	−1.469	0.000721	0.00127	0.0145***
	(0.00153)	(0.00118)	(1.393)	(0.000527)	(0.00175)	(0.00431)
Household members (count)	−0.00314	−0.00277	−12.71	−0.000165	0.00671*	0.0132
	(0.00399)	(0.00304)	(16.65)	(0.00168)	(0.00349)	(0.0128)
Value of productive assets	2.62e-09	1.13e-09	3.83e-06	1.47e-09	6.53e-10	1.05e-08
	(4.22e-09)	(3.64e-09)	(4.44e-06)	(2.42e-09)	(3.02e-09)	(1.37e-08)
Travel time to market	0.000200***	0.000125*	−0.378	−9.86e-05***	−2.06e-05	−0.000389**
	(7.40e-05)	(7.31e-05)	(0.290)	(3.12e-05)	(7.21e-05)	(0.000176)
Observations	10,009	10,009	10,009	10,009	10,009	10,009
Region FE	Yes	Yes	Yes	Yes	Yes	Yes
Year FE	Yes	Yes	Yes	Yes	Yes	Yes
Mundlak–Chamberlain device	Yes	Yes	Yes	Yes	Yes	Yes

Notes: Significance levels are denoted as follows: *= p<0.10, **= p<0.05, *** = p<0.01. FE = fixed effects.

Table 6A2. Full regression estimates for Tanzania (specification 2, see Table 6.4).

	(7)	(8)	(9)	(10)	(11)	(12)
Variables	Commercialized share of prod.	High-value share of prod.	Fertilizer (kg/ha)	Pesticide (=1)	Herbicide (=1)	Improved maize seed (=1)
Avg. age of household adults	–0.00386***	–0.00139***	–3.747*	–0.000500***	–0.00127***	–0.00359***
	(0.000349)	(0.000285)	(2.097)	(0.000165)	(0.000279)	(0.000848)
Female head (1/0)	0.0142	–0.0323*	10.96	0.00189	0.0115	–0.00516
	(0.0329)	(0.0181)	(17.83)	(0.0111)	(0.0297)	(0.0771)
Farm size (ha)	0.00393**	0.00198*	–1.555	0.000720	0.00126	0.0145***
	(0.00159)	(0.00119)	(1.392)	(0.000531)	(0.00176)	(0.00435)
Household members (count)	–0.00860**	–0.00472	–18.02	–0.000865	0.00492	0.00827
	(0.00411)	(0.00308)	(17.42)	(0.00169)	(0.00349)	(0.0127)
Value of productive assets	2.89e-09	1.23e-09	4.15e-06	1.51e-09	7.38e-10	1.08e-08
	(4.23e-09)	(3.64e-09)	(4.46e-06)	(2.42e-09)	(3.01e-09)	(1.37e-08)
Travel time to market	0.000205***	0.000125*	–0.374	–9.86e-05***	–1.98e-05	–0.000394**
	(7.55e-05)	(7.40e-05)	(0.289)	(3.12e-05)	(7.22e-05)	(0.000176)
Observations	10,009	10,009	10,009	10,009	10,009	10,009
Region FE	Yes	Yes	Yes	Yes	Yes	Yes
Year FE	Yes	Yes	Yes	Yes	Yes	Yes
Mundlak–Chamberlain	Yes	Yes	Yes	Yes	Yes	Yes

Notes: Significance levels are denoted as follows: * = p<0.10, ** = p<0.05, *** = p<0.01. FE = fixed effects.

Table 6A3. Full regression estimates for Tanzania (specification 3, see Table 6.4).

	(13)	(14)	(15)	(16)	(17)	(18)
Variables	Commercialized share of prod.	High-value share of prod.	Fertilizer (kg/ha)	Pesticide (=1)	Herbicide (=1)	Improved maize seed (=1)
Age of household head (years)	−0.00333***	−0.000811***	−1.841	−0.000267*	−0.000738***	−0.000800
	(0.000336)	(0.000280)	(1.223)	(0.000159)	(0.000240)	(0.000849)
Avg. age of other household adults	−0.000705*	−0.000620*	−0.484	4.94e-05	−0.000545*	−0.00209*
	(0.000406)	(0.000318)	(0.867)	(0.000222)	(0.000327)	(0.00113)
Female head (1/0)	0.0266	−0.0456**	19.86	0.00965	0.0222	0.00101
	(0.0393)	(0.0208)	(18.71)	(0.0132)	(0.0250)	(0.0845)
Farm size (ha)	0.00378**	0.00211*	−1.641	0.000656	0.00122	0.0144***
	(0.00150)	(0.00118)	(1.358)	(0.000515)	(0.00174)	(0.00424)
Household members (count)	−0.00492	−0.00423	−14.02	−0.000832	0.00616	0.0156
	(0.00438)	(0.00315)	(17.70)	(0.00185)	(0.00381)	(0.0135)
Value of productive assets	2.52e-09	9.61e-10	5.05e-06	1.55e-09	6.60e-10	9.53e-09
	(4.26e-09)	(3.66e-09)	(4.67e-06)	(2.43e-09)	(3.05e-09)	(1.37e-08)
Travel time to market	0.000219***	0.000143**	−0.146	−0.000101***	−1.99e-05	−0.000407**
	(7.43e-05)	(7.18e-05)	(0.169)	(3.38e-05)	(7.40e-05)	(0.000187)
Observations	9048	9048	9048	9048	9048	9048
Region FE	Yes	Yes	Yes	Yes	Yes	Yes
Year FE	Yes	Yes	Yes	Yes	Yes	Yes
Mundlak–Chamberlain	Yes	Yes	Yes	Yes	Yes	Yes

Notes: Significance levels are denoted as follows: * = p<0.10, ** = p<0.05, *** = p<0.01. FE = fixed effects.

Table 6B1. Full regression estimates for Zambia (specification 1, see Table 6.4).

	(1)	(2)	(3)	(4)	(5)	(6)
Variables	Commercialized share of prod.	High-value share of prod.	Fertilizer (kg/ha)	Pesticide (=1)	Herbicide (=1)	Improved maize seed (=1)
Age of household head (years)	–0.00153***	–0.000682***	–0.220***	–0.000386***	–0.000123	–0.000629*
	(0.000194)	(0.000174)	(0.0826)	(9.96e-05)	(9.53e-05)	(0.000373)
Max. educational attainment	0.0119***	–0.00309***	10.54***	–0.000673	0.00412***	0.0309***
	(0.00120)	(0.000905)	(0.617)	(0.000483)	(0.000748)	(0.00169)
Female head (1/0)	–0.0426***	–0.0186**	–2.689	–0.0181***	–0.00416	–0.0602***
	(0.00761)	(0.00761)	(3.518)	(0.00355)	(0.00364)	(0.0144)
Farm size (ha)	0.00142**	0.00113**	–0.755***	0.000152	0.000451*	0.000120
	(0.000623)	(0.000443)	(0.201)	(0.000142)	(0.000249)	(0.000694)
Household members (count)	0.000133	0.00748***	–2.685**	0.000287	0.000871	–0.00307
	(0.00262)	(0.00247)	(1.240)	(0.00128)	(0.00195)	(0.00472)
Productive assets (ZMW)	3.68e-08	3.28e-08	–6.97e-05***	3.74e-08	1.02e-07**	–3.75e-08
	(3.18e-08)	(2.02e-08)	(2.46e-05)	(2.99e-08)	(4.95e-08)	(5.08e-08)
Hours to town of 50k+	9.04e-07	–8.02e-06	–0.0166***	7.01e-06*	1.19e-05***	–5.59e-05***
	(7.37e-06)	(6.25e-06)	(0.00424)	(3.89e-06)	(3.26e-06)	(1.52e-05)
Population density	0.184***	0.0738**	54.85***	0.0529***	0.0353***	0.341***
	(0.0216)	(0.0330)	(8.538)	(0.0157)	(0.0134)	(0.0186)
Rainfall (mm)	0.000230***	8.03e-05**	0.0236**	1.19e-05	–1.71e-05	0.000127**
	(2.96e-05)	(3.59e-05)	(0.0118)	(1.19e-05)	(1.34e-05)	(5.67e-05)
Intraseasonal rainfall CV	0.0136	0.259***	11.38	0.0781***	–0.0532	–0.0434
	(0.0581)	(0.0577)	(20.58)	(0.0264)	(0.0324)	(0.106)
Observations	16,273	16,273	16,273	16,273	16,273	16,273
District FE	Yes	Yes	Yes	Yes	Yes	Yes
Year FE	Yes	Yes	Yes	Yes	Yes	Yes
Mundlak–Chamberlain	Yes	Yes	Yes	Yes	Yes	Yes

Notes: Significance levels are denoted as follows: * = p<0.10, ** = p<0.05, *** = p<0.01. FE = fixed effects.

Table 6B2. Full regression estimates for Zambia (specification 2, see Table 6.4).

	(7)	(8)	(9)	(10)	(11)	(12)
Variables	Commercialized share of prod.	High-value share of prod.	Fertilizer (kg/ha)	Pesticide (=1)	Herbicide (=1)	Improved maize seed (=1)
Avg. age of household adults	−0.00164***	−0.000966***	0.0655	−0.000655***	5.65e-05	−0.000292
	(0.000270)	(0.000241)	(0.117)	(0.000141)	(0.000122)	(0.000488)
Max. educational attainment	0.0109***	−0.00369***	10.60***	−0.00109**	0.00418***	0.0308***
	(0.00124)	(0.000930)	(0.632)	(0.000513)	(0.000776)	(0.00172)
Female head (1/0)	−0.0505***	−0.0213***	−4.906	−0.0198***	−0.00542	−0.0672***
	(0.00749)	(0.00744)	(3.447)	(0.00354)	(0.00363)	(0.0142)
Farm size (ha)	0.00158**	0.00116***	−0.710***	0.000175	0.000377*	0.000170
	(0.000620)	(0.000442)	(0.194)	(0.000135)	(0.000208)	(0.000663)
Household members (count)	−0.00180	0.00632***	−3.170**	−0.000849	0.000704	−0.00511
	(0.00255)	(0.00244)	(1.245)	(0.00126)	(0.00189)	(0.00468)
Productive assets (ZMW)	3.76e-08	3.45e-08*	−7.01e-05***	3.78e-08	1.01e-07**	−3.76e-08
	(3.13e-08)	(1.97e-08)	(2.46e-05)	(3.00e-08)	(4.89e-08)	(5.07e-08)
Hours to town of 50k+	7.76e-07	−8.16e-06	−0.0162***	6.65e-06*	1.18e-05***	−5.51e-05***
	(7.41e-06)	(6.30e-06)	(0.00419)	(3.96e-06)	(3.28e-06)	(1.51e-05)
Population density	0.185***	0.0743**	54.67***	0.0536***	0.0351***	0.341***
	(0.0215)	(0.0329)	(8.540)	(0.0156)	(0.0134)	(0.0188)
Rainfall (mm)	0.000235***	7.98e-05**	0.0251**	1.37e-05	−1.54e-05	0.000127**
	(2.91e-05)	(3.54e-05)	(0.0121)	(1.18e-05)	(1.32e-05)	(5.45e-05)
Intraseasonal rainfall CV	0.0133	0.256***	13.94	0.0777***	−0.0541*	−0.0374
	(0.0571)	(0.0579)	(22.02)	(0.0265)	(0.0325)	(0.105)
Observations	16,433	16,433	16,433	16,433	16,433	16,433
District FE	Yes	Yes	Yes	Yes	Yes	Yes
Year FE	Yes	Yes	Yes	Yes	Yes	Yes
Mundlak–Chamberlain	Yes	Yes	Yes	Yes	Yes	Yes

Notes: Significance levels are denoted as follows: * = p<0.10, ** = p<0.05, *** = p<0.01. FE = fixed effects.

Table 6B3. Full regression estimates for Zambia (specification 3, see Table 6.4).

Variables	(13) Commercialized share of prod.	(14) High-value share of prod.	(15) Fertilizer (kg/ha)	(16) Pesticide (=1)	(17) Herbicide (=1)	(18) Improved maize seed (=1)
Age of household head (years)	−0.00141***	−0.000454**	−0.416***	−0.000224*	−0.000289**	−0.00111**
	(0.000233)	(0.000195)	(0.109)	(0.000118)	(0.000137)	(0.000446)
Avg. age of other household adults	−0.000261	−0.000141	0.352**	−0.000336*	0.000452**	0.000636
	(0.000339)	(0.000278)	(0.158)	(0.000175)	(0.000205)	(0.000551)
Max. educational attainment	0.0119***	−0.00315***	11.38***	−0.000889	0.00488***	0.0321***
	(0.00133)	(0.00101)	(0.661)	(0.000564)	(0.000880)	(0.00178)
Female head (1/0)	−0.0505***	−0.0185**	−4.759	−0.0193***	−0.00311	−0.0569***
	(0.00869)	(0.00759)	(4.293)	(0.00431)	(0.00426)	(0.0165)
Farm size (ha)	0.00166***	0.00110**	−0.635***	0.000155	0.000431*	8.28e-05
	(0.000581)	(0.000455)	(0.190)	(0.000159)	(0.000238)	(0.000761)
Household members (count)	0.000689	0.00647**	−1.454	−0.000271	0.00116	−0.00206
	(0.00281)	(0.00267)	(1.428)	(0.00143)	(0.00199)	(0.00516)
Productive assets (ZMW)	3.13e-08	3.30e-08*	−7.39e-05***	4.05e-08	9.19e-08**	−4.17e-08
	(3.33e-08)	(2.00e-08)	(2.49e-05)	(3.16e-08)	(4.45e-08)	(5.09e-08)
Hours to town of 50k+	−4.07e-06	−5.96e-06	−0.0181***	7.25e-06*	1.11e-05***	−6.62e-05***
	(7.59e-06)	(6.14e-06)	(0.00444)	(3.90e-06)	(3.14e-06)	(1.56e-05)
Population density	0.189***	0.0729**	61.59***	0.0586***	0.0391***	0.355***
	(0.0224)	(0.0296)	(9.164)	(0.0165)	(0.0145)	(0.0175)
Rainfall (mm)	0.000234***	8.67e-05**	0.0202	1.33e-05	−1.96e-05	0.000112*
	(3.08e-05)	(3.39e-05)	(0.0129)	(1.26e-05)	(1.47e-05)	(6.04e-05)
Intraseasonal rainfall CV	0.00327	0.279***	11.33	0.0857***	−0.0528	−0.0165
	(0.0614)	(0.0552)	(22.26)	(0.0273)	(0.0346)	(0.112)
Observations	15,178	15,178	15,178	15,178	15,178	15,178
District FE	Yes	Yes	Yes	Yes	Yes	Yes
Year FE	Yes	Yes	Yes	Yes	Yes	Yes
Mundlak–Chamberlain	Yes	Yes	Yes	Yes	Yes	Yes

Notes: Significance levels are denoted as follows: * = p<0.10, ** = p<0.05, *** = p<0.01. FE = fixed effects.

Table 6C1. Full regression estimates for Ethiopia (specification 1, see Table 6.4).

	(1)	(2)	(3)	(4)	(5)	(6)
Variables	Commercialized share of prod.	High-value share of prod.	Fertilizer (kg/ha)	Pesticide (=1)	Herbicide (=1)	Improved maize seed (=1)
Age of household head (years)	–0.000519*	5.16e-05	0.000121	–0.000401	–0.00133**	0.000213
	(0.000277)	(0.000266)	(0.000403)	(0.000393)	(0.000564)	(0.000528)
Max. educational attainment	0.000489	0.00249	0.00656***	–0.00277	0.00370	0.00300
	(0.00253)	(0.00245)	(0.00253)	(0.00286)	(0.00321)	(0.00290)
Female head (1/0)	–0.0374	–0.0110	0.0241	–0.0500	–0.0213	0.0315
	(0.0263)	(0.0328)	(0.0398)	(0.0306)	(0.0444)	(0.0278)
Farm size (ha)	0.000164	0.000470	0.000967**	–0.000531	0.00140	0.000125
	(0.000285)	(0.000739)	(0.000376)	(0.000453)	(0.00122)	(0.000284)
Household members (count)	0.00422	–0.0122	0.0107	0.00235	–0.00928	0.00767
	(0.00619)	(0.00771)	(0.00751)	(0.00821)	(0.00780)	(0.00709)
Travel time to market	–1.72e-05	5.58e-05	–0.000566***	–0.000205**	–0.000343*	–0.000558***
	(6.59e-05)	(5.72e-05)	(0.000123)	(9.20e-05)	(0.000205)	(0.000177)
Observations	9285	9285	9285	9285	9285	9285
Regional FE	Yes	Yes	Yes	Yes	Yes	Yes
Year FE	Yes	Yes	Yes	Yes	Yes	Yes
Mundlak–Chamberlain	Yes	Yes	Yes	Yes	Yes	Yes

Notes: Significance levels are denoted as follows: * = p<0.10, ** = p<0.05, *** = p<0.01. FE = fixed effects.

Table 6C2. Full regression estimates for Ethiopia (specification 2, see Table 6.4).

	(7)	(8)	(9)	(10)	(11)	(12)
Variables	Commercialized share of prod.	High-value share of prod.	Fertilizer (kg/ha)	Pesticide (=1)	Herbicide (=1)	Improved maize seed (=1)
Avg. age of household adults	−0.00108***	−9.64e-07	−0.000509	−0.00111**	−0.00246***	−0.000108
	(0.000407)	(0.000466)	(0.000606)	(0.000443)	(0.000817)	(0.000779)
Max. educational attainment	0.000214	0.00243	0.00634**	−0.00300	0.00323	0.00286
	(0.00253)	(0.00244)	(0.00251)	(0.00288)	(0.00320)	(0.00286)
Female head (1/0)	−0.0426	−0.0136	0.0239	−0.0514*	−0.0168	0.0288
	(0.0261)	(0.0324)	(0.0388)	(0.0299)	(0.0436)	(0.0266)
Farm size (ha)	0.000164	0.000470	0.000970**	−0.000530	0.00140	0.000129
	(0.000288)	(0.000739)	(0.000379)	(0.000457)	(0.00122)	(0.000284)
Household members (count)	0.00374	−0.0121	0.0104	0.00170	−0.0104	0.00755
	(0.00619)	(0.00766)	(0.00754)	(0.00822)	(0.00789)	(0.00716)
Travel time to market	−1.62e-05	5.53e-05	−0.000568***	−0.000205**	−0.000339*	−0.000560***
	(6.56e-05)	(5.72e-05)	(0.000123)	(9.23e-05)	(0.000205)	(0.000177)
Observations	9297	9297	9297	9297	9297	9297
Regional FE	Yes	Yes	Yes	Yes	Yes	Yes
Year FE	Yes	Yes	Yes	Yes	Yes	Yes
Mundlak–Chamberlain	Yes	Yes	Yes	Yes	Yes	Yes

Notes: Significance levels are denoted as follows: * = p<0.10, ** = p<0.05, *** = p<0.01. FE = fixed effects.

Table 6C3. Full regression estimates for Ethiopia (specification 3, see Table 6.4).

Variables	(13) Commercialized share of prod.	(14) High-value share of prod.	(15) Fertilizer (kg/ha)	(16) Pesticide (=1)	(17) Herbicide (=1)	(18) Improved maize seed (=1)
Age of household head (years)	–0.000472	1.97e-05	0.000276	–9.47e-05	–0.00103*	0.000297
	(0.000296)	(0.000279)	(0.000459)	(0.000463)	(0.000609)	(0.000614)
Avg. age of other household adults	–0.000344	0.000529	0.000461	–0.00127**	–0.00142	–0.000177
	(0.000467)	(0.000641)	(0.000671)	(0.000553)	(0.000987)	(0.000965)
Max. educational attainment	0.000157	0.00193	0.00684**	–0.00261	0.00375	0.00297
	(0.00256)	(0.00249)	(0.00266)	(0.00297)	(0.00336)	(0.00300)
Female head (1/0)	–0.0391	–0.0224	0.0310	–0.0494	–0.0104	0.0363
	(0.0269)	(0.0333)	(0.0381)	(0.0320)	(0.0468)	(0.0307)
Farm size (ha)	0.000199	0.000540	0.000886**	–0.000554	0.00147	0.000140
	(0.000276)	(0.000835)	(0.000415)	(0.000457)	(0.00129)	(0.000304)
Household members (count)	0.00578	–0.0134	0.00823	0.00662	–0.0118	0.00744
	(0.00645)	(0.00815)	(0.00741)	(0.00856)	(0.00823)	(0.00750)
Travel time to market	–2.43e-05	6.38e-05	–0.000530***	–0.000220**	–0.000349	–0.000584***
	(6.97e-05)	(6.06e-05)	(0.000125)	(9.55e-05)	(0.000219)	(0.000187)
Observations	8523	8523	8523	8523	8523	8523
Regional FE	Yes	Yes	Yes	Yes	Yes	Yes
Year FE	Yes	Yes	Yes	Yes	Yes	Yes
Mundlak–Chamberlain	Yes	Yes	Yes	Yes	Yes	Yes

Notes: Significance levels are denoted as follows: * = p<0.10, ** = p<0.05, *** = p<0.01. FE = fixed effects.

Table 6D. Factor-based farm orientation measures for Zambia (see Table 6.6).

Variables	Factor 1 (intensification type A)	Factor 2 (intensification type B)
Commercialized share of production	0.2738	0.2317
High-value share of production	–0.0582	0.4112
Fertilizer (kg/ha)	0.3737	–0.0641
Pesticide (farm share)	–0.0323	0.4177
Herbicide (farm share)	0.0690	0.0537
Improved maize seed (=1)	0.4139	–0.0541

Notes: Table shows scoring coefficients based on varimax rotated factors. Factors are derived from variables measuring: commercialized share of production; high-value share of crop production; fertilizer application rate (kg/ha); share of land under pesticide; share of land under herbicide; and usage of improved maize seed.

7 The Social Landscape of Education and Work in Rural Sub-Saharan Africa

Barbara Crossouard[1], Máiréad Dunne[1] and Carolina Szyp[2]
[1]University of Sussex, Brighton, UK; [2]Institute of Development Studies, Brighton, UK

Introduction

Education has long been constructed as intrinsic to the development and modernization of sub-Saharan African (SSA). A dominant rationale for investment in education in SSA has framed it predominantly in terms of the development of human capital (Schultz, 1961; Becker, 1962; Montenegro and Patrinos, 2014). With reference to the arguments of this chapter, a key issue is that human capital depends on the separation of education and work, as if these are clearly distinct and separable activities, whose relationship can be analysed as a closed system. In addition, as we discuss below, the macroeconomic perspectives of human capital theory (HCT) provide few insights into the local-level interactions between education and work, or the historical educational inequalities that have prevailed in and across SSA. This chapter draws on qualitative research into youth livelihoods in four SSA countries that has addressed the local social dynamics of work and education from the perspectives of young people themselves. Firstly, it illuminates the extent to which youth in the four different national contexts value education. It then turns to young people's lived experiences of juggling both schooling and work from an early age, highlighting the wide disparity between idealized notions of 'transition' and the complexities of youth livelihoods. Finally, it explores the gendered dimensions of this social landscape, and how these produce different pressures that force young women in particular out of education. We conclude with implications for young people's current and future engagement with the rural economy, and for education policy.

Education and Work

This section locates our analysis within the policy and research literatures. It first gives an overview of policy concerns about the education of young women and men in SSA. Noting that the research discussed here spans Sustainable Development Goal 4 (SDG 4) (education quality), SDG 5 (gender equality) and SDG 8 (the right to decent work), we engage in particular with HCT, as the dominant lens used within international policy to understand the relationship between education and work. We challenge HCT's binary construction of education and work as if these were mutually exclusive. We critique how the theory's universalistic assumptions lead to the construction of individualized deficit, where limited engagement in education by young people in poor rural contexts is attributed to their lack of aspiration. This contrasts sharply with our analysis, which shows that rural youth value education highly.

DOI: 10.1079/9781789245011.0007

We further critique HCT's assumptions of a linear construction of education and work as if these were two distinct life stages. In particular, the reduction of the 'education–work' relationship to a 'black box' simply fails to attend to the complexities of young people's engagement in the nexus of work and education within the informal economies of rural SSA. Importantly, for the final strand of our analysis, the way HCT conceptualizes work – and in particular the privileging of wage employment – compounds the historical misrecognition of women's invisible work and fails to address the gendered economy of education and work in these contexts.

Having argued for the significance of context, we conclude this section by turning to research which illuminates the elitist and gendered ways education was developed in SSA in colonial times, how this has associated education with social mobility, and set in motion particular imaginaries of both education and work (see also Chapter 9, this volume). Drawing upon contemporary research into education and work, we highlight how these legacies endure, so that the processes of schooling and work continue to reproduce inequalities and exclusions, in which gender is routinely implicated.

Education inequalities

Education has long been a key concern of international policy agendas, such as Education for All, the Millennium Development Goals and the Sustainable Development Goals. A strong policy focus on increased access resulted in enrolment in primary education in SSA rising by 75% between 1999 and 2012 (UNESCO, 2015). Despite the recent shift in focus to educational quality (The World Bank, 2018; Wulff, 2020), access to learning remains a critical issue that cannot be considered separate from a concern for quality. UNESCO (2018) show that out-of-school rates in SSA increase throughout schooling, rising from 21% at primary level, to 36% at lower secondary and 57% at upper secondary.

When further disaggregated, these statistics illuminate deep inequalities. For example, while the adjusted gender parity index for 2017 shows slightly more females completing primary school, by upper secondary fewer than four females complete school for every five males (UNESCO, 2018). SSA also has the highest level of out-of-school youth of any region of the world (31%), although this is much worse in rural contexts (37%) than urban contexts (20%). Differences between rich and poor are even more stark, with 51% of the poorest youth being out of school versus 16% of the richest. In addition, SSA has the lowest percentages of any world region for trained primary and secondary school teachers. Overall, what these figures show is that education in SSA is riven by inequalities that continue to benefit urban populations rather than rural, richer people rather than poorer, and males rather than females.

As noted above, the overriding policy concern has now veered towards educational quality, with urgent calls being voiced by institutions such as The World Bank to address what they call a 'global learning crisis' (The World Bank, 2018). In the eyes of some powerful actors, this provides further justification for the deregulation, privatization and liberalization of the education sector. The introduction of market dynamics into education (Benavot and Smith, 2020) continues what was started by the Structural Adjustment Programs (SAPs) of the International Monetary Fund (IMF) and The World Bank. The implementation of SAPs since the 1980s has been strongly critiqued, particularly for their impact on state spending for social infrastructure, including education (Jomo and Fine, 2006; Tikly, 2019). As Ilon (1994) comments, the resulting hollowing out of the state created the conditions for the rise of low-fee private schooling in many Global South contexts, including SSA. Diminished state funds to support public education – particularly at a time of strong demographic growth – led to a middle-class flight from state schools, which has contributed to the difficulties of sustaining the quality of provision. Importantly, as discussed further below, current policy concerns about quality – and particularly how it should be quantified – have reduced education to 'learning' in ways that have diverted attention from the politics of education (Hossain and Hickey, 2019).

Policy perspectives on education and work

As Marginson (2019) notes, policy narratives that connect education and work have demonstrated an unquestioning reliance on HCT (see, e.g. The World Bank, 2020). HCT frames investment in education in terms of the development of human capital, which translates in turn into improved life opportunities and higher wages for the individual, and more widely, to increased productivity, consumption and economic development. The key metric used in HCT is rate of return to schooling, which involves calculating the proportional increase in an individual's labour market earnings from each additional year of schooling completed. Montenegro and Patrinos (2014) provide further elaboration, including details of the statistical processes.

Many critiques of HCT have been voiced. These include questions about its evidence base; its assumptions about opportunity costs (Bennell, 1996); its static economic model (Resnik, 2006); its generalization of a particular set of socio-economic relations to other quite different economic and educational contexts (Marginson, 2019); and relatedly, its failure to take account of the different histories of education systems in the Global South (McGrath *et al.*, 2019). Overall, HCT directly correlates investment in education with productivity increases and higher work-related earnings. However, this linear relationship does not account for education as a complex of socially situated processes. As Resnik (2006) puts it, HCT has become a *doxa*, a set of unquestioned presuppositions, that reductively assumes an 'education–economic growth black box'.

The black box assumes linear, incremental relations between education and work in a developmental trajectory that is both individual and societal. For the individual, it involves a normative journey through childhood, youth and finally to adulthood. Education is critical to this trajectory, preparing individuals for the world of work as part of a step by step progression through different life stages. For instance, The World Bank (2006) assumes the life course to reflect 'five phases' of individual development, with learning being the initial stage, followed by 'starting a productive working life', 'adopting a healthy lifestyle', 'forming a family', and 'exercising citizenship'. These phases are projected as if they are near universal. As we show below, in the contexts of our research it is impossible to divide the life course into neat phases, where learning necessarily comes before starting to work; where starting a family is distinct from schooling; or where having responsibilities is something that can be associated with adulthood.

Despite the push to engage children in schooling, the gaps in uptake have been of some concern, particularly in low-income contexts which would seem to have most to gain from investment in education, whether for the individual or more widely in terms of national development. The World Bank (2014) discusses how poverty creates a 'failure of aspiration' that results in (and from) limited engagement in schooling. Here, the explanation for non-engagement in schooling is readily turned into a question of individual deficit, related to individuals' failure to recognize the (economic) benefits of education. Although engaging briefly with more sociological theories, this document from The World Bank highlights as key concerns the 'choices' made by poor people and the 'irrationality' of their beliefs. This implicitly takes up the language of rational actor theory, whose 'narrow economistic' understanding of practice lacks attention to context and ignores the collective histories through which particular 'structures of preference' have been constituted (Bourdieu and Wacquant, 1992, p.123).

Similarly, a World Bank analysis of the high number of out-of-school youth in SSA suggests that this problem can be addressed by changing the attitudes to education of young people and parents (Inoue *et al.*, 2015). Drawing on psychological research, much of it conducted outside of SSA, these authors pinpoint young people's 'behaviour and personality traits, goals, motivations and preferences' (Inoue *et al.*, 2015, p.48) as leading to poor educational outcomes. Again, such commentary constructs the problem as an issue of individual deficit, and context is assumed irrelevant. When this report considers more contextualized research (e.g. Pryor and Ampiah, 2003), the complex range of issues this illuminates (including concerns about education quality) are again used to reiterate the dominant trope that poor attitudes to schooling are the default cause of low educational participation and poor outcomes. As more research explores rural youth livelihoods in the Global South, this trope is being increasingly challenged, however, including by the analysis we present below.

Turning more directly to work, a further limitation of HCT is the way it privileges wage employment in the formal economy. This may still be a dominant mode of economic engagement in the Global North, but it is very far from accounting for the livelihood experiences of people – young and old – in rural contexts of SSA. In this region there is a large informal economy[i], in which both young people and women are over-represented (see also Sumberg *et al.*, 2020). The International Labour Organization (ILO, 2018b) reports for example that 85% of employment in Africa is in the informal sector, with informal sector employment accounting for almost 90% of women's employment, as opposed to 82% of men's employment. For young people, this increases to almost 95%. Significantly for our interests here, informal employment dominates agriculture, at almost 98%. ILO (2018b) notes that globally, informal employment is closely related to education levels; in Africa, those with no education are predominantly engaged in informal employment (94%); reducing to 88% for those with primary education; 68% for those with secondary education; and finally, to 27% for those with tertiary education.

Informal economies also include high levels of unpaid work and this falls disproportionately on women (McDowell, 2014; ILO, 2017; UN Women and ILO, 2020). As Finlay *et al.* (2019) point out, the Living Standards Measurement Survey – Integrated Surveys on Agriculture (LSMS–ISA), overseen by The World Bank, positions women's work as secondary to that of men, and by privileging regular, paid activity, makes it likely that women's work is not recognized as 'work' at all. Importantly, women in Africa are over-represented in 'vulnerable'[ii] employment categories. ILO (2018a) reports that 76% of women are in such work, compared to 58% of men.

Drawing on LSMS–ISA data for Uganda, Koolwal (2019) shows that rural women have the highest total work burdens of paid and unpaid work compared to other groups (rural men, urban men and urban women). She highlights the need for more research into how to record rural women's work, particularly as the definition of employment has recently been narrowed to 'work for pay or profit' by the International Conference of Labour Statisticians. This will compound the invisibility of women's reproductive work (e.g. caring for younger siblings, water and fuel collection, cooking, etc.), which historically has been misrecognized, even if it has always been central to capitalist economies (Pateman, 1988; Butler, 1997; Mies, 2014). As Rai *et al.* (2019) note, while SDG 5 (focused on gender equality) calls for the recognition of women's reproductive and domestic labour, SDG 8 (focused on the right to decent work) continues to use indicators that prioritize work in the formal economy. Importantly, all these measures conflate sex and gender, and in so doing, reinscribe a female/male binary (Dunne, 2008).

Contextualizing education and work

The strong association between informal work and lower levels of education is important in relation to the promise of social mobility that education is seen to offer. Here, the history of schooling in SSA is significant. Its introduction during colonial eras was far from universal – schools were developed first in urban areas and generally served only elite groups, on a single-sex basis. Schooling was strongly influenced by missionary activity, in ways that were gendered and racialized (Clignet and Foster, 1964; Scanlon, 1964; Bolibaugh, 1972; Ball, 1983; McClintock, 1995; Vavrus, 2002; Leach, 2008; Healy, 2011; Guidi, 2018; Bryant, 2020; Crossouard and Dunne, 2020). Rural areas were largely neglected, while the selection of knowledge and its organization in school curricula generally followed what was done in the colonizing countries, and served the colonial project (Dunne and Adzahlie-Mensah, 2016; Feldman, 2016; Adzahlie-Mensah and Dunne, 2018; Bryant, 2020). The symbolic authority of western schooling came to be seen as a passport to a secure professional career and social mobility (Ball, 1983; Stockwell, 2012). Moreover, the differential provision of education sedimented social divisions between the urban and the rural that endure to this day (Nwauwa, 2020).

Largely, all SSA countries now attempting to work towards SDG 4 (foreground education quality) and SDG 5 (gender equality) face multiple challenges. Research has shown that legacies of colonialism have endured in schools in ways that reproduce inequities and exclusions (Dunne and Adzahlie-Mensah, 2016; Adzahlie-Mensah and Dunne, 2018). In addition to exclusions related to location, religion, language of instruction and ethnicity, this literature shows the practices of schooling to be deeply gendered. Indeed, drawing upon theoretical insights from

Butler (1990) and Connell (2005), it demonstrates how the institutionalized practices of schooling serve to secure the reproduction of gender binaries in ways that can involve gender violence (Mirembe and Davies, 2001; Dunne *et al.*, 2005, 2006; Bakari and Leach, 2007; Dunne, 2007, 2008; Humphreys *et al.*, 2015).

Alongside literature focused on gender and education in SSA, some research into youth livelihoods takes up the intersections of schooling and work in different rural contexts of the Global South. This generally suggests that even in the face of considerable obstacles, young people value education, seeing it as key to realizing social mobility and securing a professional career (see also Camfield, 2011; Boyden, 2013; Yeboah *et al.*, 2017; Ansell, 2018). Like Farrugia (2018), this literature often problematizes the notion of 'transition' from school to work, and then to independent adulthood, showing this to be a western construction that bears little resemblance to youth livelihoods in SSA (see also Nilsson, 2019). Rather than a linear, sequential transition from school to work, to parenthood, synchronous working and schooling is the norm. This includes paid and unpaid work, which together often provide rural families with little more than basic subsistence (Boyden, 2013; Chuta and Morrow, 2015; Maconachie and Hilson, 2016). Some young people make valiant attempts to continue their schooling, but navigating the very different spaces of schooling and their complex lives in the community leads to significant identity tensions. For example, young people who have considerable social responsibilities, as parents, carers and/or wage earners, can find themselves infantilized and humiliated by their treatment in the classroom (Dunne and Ananga, 2013).

Research Contexts

We now turn to our research into rural youth livelihoods and imagined futures in four SSA countries: Côte d'Ivoire and Nigeria in West Africa; and Ethiopia and Uganda in East Africa. These countries all have large youth populations, large rural economies, low educational outcomes, and important gender gaps in educational access and outcomes. All four have a history of state turbulence, fuelled by ethnic divisions that were sedimented during the age of imperialism. Nigeria and Uganda were once British colonies, and Côte d'Ivoire was part of French West Africa. Ethiopia was not formally colonized, but still witnessed interventions reflecting the imperial ambitions of different European countries. To frame the discussion, Table 7.1 below draws on the World Inequality Database on Education (WIDE). The figures presented illustrate the poor educational outcomes for young people across the four different countries, in addition to important differences for females and males, and between rural and urban contexts. It is important to note that these statistical aggregates conceal the very different populations of these countries. We also note the fallibility of such data, including for example that participation data are likely to be exaggerated (Humphreys *et al.*, 2015). A further caution is that national-level data conceals wide disparities between regions within a country. In Nigeria, for example, the study did not include any sites in the north, where education outcomes are significantly worse than in the south.

Despite their limitations, these national statistics indicate considerable differences between rural and urban contexts across all measures. For instance, with respect to primary completion, those individuals in the rural contexts of Ethiopia are less than half as likely to complete than urban pupils. Gender differences are evident – while female and male completion rates at primary school are at parity in some contexts, in all four countries female youth are more likely to be out of school than male youth: just over half of female youth are out of school in Nigeria and Ethiopia, increasing to 72% in Uganda. Intersecting gender and rural/urban categories would show yet deeper differences in favour of males.

Table 7.2 draws on LSMS–ISA data and depicts labour allocations for female and male youth, again confirming differences in young females' engagement in schooling, and important differences in the extent females and males are involved in farming, the wage employment, or report no economic activity. As noted earlier, an important qualification here is how LSMS–ISA methodologies can lead to under-reporting of women's work. For example, Koolwal (2019) discusses the high levels of 'own use' agricultural production in Nigerian survey data, particularly for women, noting that under

Table 7.1. Selected education statistics for Uganda, Ethiopia, Nigeria and Côte d'Ivoire. From World Inequality Database on Education (WIDE, www.education-inequalities.org/, accessed 26 November 2020). For definitions of indicator categories, see www.education-inequalities.org/indicators.

		Gender		Location	
Country / measure	National average	Female	Male	Rural	Urban
Uganda					
Primary completion (%)	49	49	49	41	77
Out-of-school children (%)	13	13	13	14	10
Upper sec. completion (%)	17	11	17	7	34
Out-of-school youth (%)	65	72	57	66	61
Higher educ. attendance (%)	5	5	6	2	15
Ethiopia					
Primary completion (%)	42	39	44	31	76
Out-of-school children (%)	32	32	32	34	17
Upper sec. completion (%)	10	8	12	4	28
Out-of-school youth (%)	49	53	44	53	35
Higher educ. attendance (%)	5	5	6	1	19
Nigeria					
Primary completion (%)	69	62	77	55	89
Out-of-school children (%)	37	37	33	44	15
Upper sec. completion (%)	44	36	67	28	67
Out-of-school youth (%)	51	54	47	58	38
Higher educ. attendance (%)	7	7	10	3	15
Côte d'Ivoire					
Primary completion (%)	51	44	59	33	65
Out-of-school children (%)	23	26	20	29	15
Upper sec. completion (%)	15	12	19	4	24
Out-of-school youth (%)	57	67	47	70	48
Higher educ. attendance (%)	4	3	4	0	6

the new definition of 'employment', this would no longer be categorized as 'self-employed in agriculture'. She also notes problems with the language of some survey questions, and whether women recognize how this relates to their work. For example, whether they would identify as being involved in 'business', when their activities are small scale and often conducted alongside household activities. The same issue of non-identification with particular categories may well apply to questions that probe time spent on particular activities. Overall, Koolwal highlights the need for more nuanced attention to contributions of women's work to the rural economy, to ensure this is adequately recognized.

Research Methodology

The study involved engagement with female and male youth using predominantly qualitative participatory research methods (see Chapter 2, this volume). These sought to develop an in-depth understanding of youth perspectives on their livelihoods and imagined futures; what part education played in both; and how all of this was gendered. The study was conducted in four different rural locations in each country. With slight variations from site to site, the research in each location involved four sex-segregated focus group discussions (FGDs), two with female youth and two with male youth; eight livelihood interviews; 12 life history interviews; six photo–voice interviews focusing more specifically on imagined futures; and six interviews with adults (see Chapter 2, this volume). Each set of interviews involved approximately equal numbers of females and males with different educational backgrounds. The interviews were conducted with the support of local researchers in each location and with participants' consent, were recorded, translated and

Table 7.2. Labour allocations of individual youth (aged 15–24) for Uganda, Ethiopia and Nigeria. From Living Standards Measurement Survey – Integrated Surveys on Agriculture (LSMS–ISA; no data for Côte d'Ivoire).

	Gender	
Country / labour allocation	Female	Male
Uganda		
Wage employment (%)	13	25
Non-farm business (%)	10	8
Farming (%)	76	78
In school (%)	47	55
No activity reported (%)	6	2
Ethiopia		
Wage employment (%)	9	7
Non-farm business (%)	12	10
Farming (%)	36	55
In school (%)	40	46
No activity reported (%)	26	18
Nigeria		
Wage employment (%)	3	3
Non-farm business (%)	11	8
Farming (%)	15	28
In school (%)	48	53
No activity reported (%)	31	20

fully transcribed. The excerpts presented here draw on this large data set.

The data were analysed using a combination of qualitative and quantitative approaches. Alongside thematic analysis, some coding was applied to provide a quantitative depiction of whether farming, migration or education figured in the interviewees' imagined futures. The analysis takes up three broad themes: first, how youth value education; second, the non-linear relationships between education and work; and finally, the gendered dimensions of these education–work relationships.

Research Findings

The value of education to rural youth

The first key point is that the majority of rural youth in our research appear to value education. This finding generally challenges accounts in policy and some research literature (psychological research in particular) that there is a need to address poor 'attitudes' to education. Youth sometimes commented on the lack of relevance of the school curriculum to rural livelihoods, declaring more interest in vocational education. However, on the whole, in addition to the benefits that literacy could bring, schooling was associated with social mobility and increased social status:

> Schooling assists one to rise to a lofty height. Education makes one professional like lawyer, doctor and other enviable jobs in the society. As well as that, schooling makes reading and writing easy.
>
> (23-year-old male, Igbariam Eziafor, Nigeria)

> When you are a teacher, people will respect you in the community, the kids you are bringing up will also respect you and you will have many friends. You will be able to give people knowledge. Also, when you are morally upright, people will admire you and try to live the way you do.
>
> (20-year-old female, Mucwini, Uganda)

> [...] Educated people have a lot of opportunities. For instance, you can't be employed by the government even in the lowest position without some level of education. You can't be a *kebele* [the smallest administrative unit] manager or employed elsewhere without education. In the case of women, there is a big difference between educated and uneducated women in terms of their physical neatness and ideas.
>
> (25-year-old male, Bora, Ethiopia)

Education often figured in youths' imagined futures (see Chapter 9, this volume), as part of a portfolio of activities combining farming with other commerce and trading, different vocational activities (e.g. welding, carpentry, hairdressing, tailoring), or professional positions. Many individuals, both females and males, expressed the desire to further their education should they have the opportunity. Where education was no longer possible for young people who were already married, it was often desired for their children, where again, the association between education and upward economic and social mobility was clear:

> I want to be successful and become a notable personality in the community. I want to build a house, ride on my car, marry a good wife and she will give birth to good children and I will train them in school and they will study what I could not study and be greater than me.
>
> (25-year-old male, Igbariam Eziafor, Nigeria)

> I didn't go to school, so why do I send my children to school? Because it is painful for me that I wasn't able to do that – it really upsets me. So that's why I say to my children that they will study, so that tomorrow they can do well in life.
> (28-year-old male, Daloa, Côte d'Ivoire)

> I would like also to raise my young children, my brother and my sister's children to study and be educated and that in the future they can get jobs.
> (male, Awach, Uganda)

This does not mean that the young people necessarily voiced a concrete plan of action for their future education, which would demand a particular set of capitals, and relatedly, an awareness of the field and how to play it (Bourdieu, 1977). Their desire to continue in education was often expressed in vague ways, sometimes with qualifiers about funding becoming available, such as 'if I can find a sponsor'. Two different coding processes for the extent that education figured in the imagined futures of the participants revealed a significant gap between those who could elaborate specific plans for their future education, as opposed to a wider group that also included those who expressed a more general sense that education was valuable, either for themselves or for their children. In the case of Nigeria, 80% of the participants valued education, although only 50% had specific plans. For Ethiopia, 82% valued education, while 41% had specific plans. Even if unformulated and perhaps unrealizable, the ambitions of those individuals without specific plans should not be discounted.

In general, as Nwauwa (2020) suggests, education is seen as opening up new, brighter futures, creating opportunities for social mobility. Critically, however, the jobs that are opened up in principle by schooling, are not readily available in the rural economy. The young women and men also showed awareness that completing school does not guarantee employment:

> There is no work. There are many youth who have completed their education but often walk aimlessly on the street.
> (33-year-old male, Kuyu, Ethiopia)

They further noted the importance of personal connections for securing salaried work – educational qualifications are not enough. As a 24-year-old male from Kuyu, Ethiopia, commented, 'I do not have any relative working in an office. I do not know any person who can support me to get a job.'

In many cases, however, this did not seem to diminish their convictions that they could improve their social status through education. Although many were forced out of education at an early stage, those who had completed secondary education spoke of their desire to continue into higher education. Awareness of the limitations in education provision in rural as opposed to urban contexts provoked commentary on social inequalities in access to education, which again was associated with high social standing:

> I have completed the education system that we have here which is secondary school level. I have to move out to study further, but I have financial constraints. [...] Money is required for everything; I don't have money and I am not educated as I wish. [...] people with good profession and rich men have a voice in the society, they are well respected and occupy esteemed positions.
> (20-year-old female, Idi Amu and Igbokiti, Nigeria)

As commonly recognized, access to primary or secondary education is often difficult and some participants had to travel several hours a day to get to school. Although elders often spoke of sending their children to school outside the community to access better education, the quality of teaching was relatively unchallenged by the young people themselves. Indeed, in stark contrast to the assumptions within international policy arenas that rural youth may not value education, it was striking that most seemed willing to devote scarce resources to pay for their education (and indeed had done so in the past). We consider this further in the next section, before exploring the gendered landscape of education and work.

Schooling and work in the rural economy

In contrast to the ideal of a linear trajectory from schooling into work that is projected within international policy circles, both female and male youth often worked from an early age. Many began to work before starting school, and continued to work throughout their studies, moving opportunistically between different ways of generating income. Combining school and work from an early age was routine, often including helping on the family farm and in the home (especially girls). Seasonal work was also common, both for females and males. As demonstrated in Humphreys *et al.* (2015), it is clear that the school timetable does not sit well with the exigencies of rural youth livelihoods in these contexts, even if these mismatches are not typically addressed within current concerns for 'quality' education.

Responses to a question about the challenges associated with combining work and school sometimes showed the major impact this could have on education:

> I was attending my school and supporting my grandmother during my free time. I did not fail in any of the Grades (1–9). I completed Grade 10. [...] However, I could not get a pass mark. I had a lot of household activities; I was the only one who supported my grandmother. I did not get sufficient time to study hard.
>
> (20-year-old female, Jabi Tehnan, Ethiopia)

We can see here the impact of unpaid, domestic labour on this young woman, an issue we return to below. In another example, a young man reflected on the impact on his schooling of having to devote time to work:

> [...] The biggest [challenge] I have is that there is no money, if there was money, I would have completed my schooling before now, that is by concentrating squarely on schooling. That is why I fish, so that I can support myself.
>
> (22-year-old male, Igbariam Eziafor, Nigeria)

However, many others refuted the idea that combining work and school was problematic:

> We went to farm between 5:30–7:00 am, then we got dressed for school by 7:30 am, and in the evening after school hours around 4:00 pm, I went for hawking. [...] There were no challenges as such, because if we do not go to farm in the morning and sell few things from the farm we will not have money to take to school. But it has never disturbed school.
>
> (25-year-old female, Oba Oke, Nigeria)

> When I was at school, when we didn't have class, my mother gave us a plot of land and we grew rice. Every Friday, I get my produce, I set up a table outside the house, and I sold it. I didn't have any issues, I had no problem combining school with trading.
>
> (28-year-old female, Daloa, Côte d'Ivoire)

In many cases continuing in school was only possible because of the income from such work; this allowed the costs of schooling to be covered, and provided some support to the young person's family:

> I started my own farming in the year 2012. It was only maize because by that time as I was still studying, and weeding maize was easier. The maize I was cultivating paid my school fees and I also gave some to my mother and we eat it at home.
>
> (22-year-old male, Awach, Uganda)

In some instances, working had been beneficial in terms of gaining experience in areas that school addressed in a more 'bookish' way. For example, a 28-year-old male in Daloa, Côte d'Ivoire, had been in and out of school, being involved in farming, masonry work and different forms of petty commerce. He had developed an interest in carpentry from working with his uncle in Abidjan and returned to school to complete a 'technical' Baccalaureate. As he remarked, 'That's when what I call good things started to happen.' He now works as a carpenter himself and is aiming to set up his own carpentry shop.

While this is a story of improving personal circumstances, where family connections have been enabling, many other young people recounted how continuing in school could be put at risk by unforeseeable downturns in family fortunes. Illness, injury or death of a parent or guardian were often cited as major reasons for individuals having to commit themselves to work, rather than education. As a young Ethiopian man commented:

> Because of the death of my father, life became difficult for us so I was forced to stop my education. If my father had huge tracts of land, we would have rented it out and used it as source of income but we don't have that much land, and I had to quit my education and started supporting myself and the family.
>
> (21-year-old male, Bora, Ethiopia)

The life story of a 31-year-old woman now living in Luweero, Uganda, is illustrative of the dislocated, gendered trajectories of some participants. She started primary school in Butambama because her brother was living there, and stayed through to Primary 4 Term 2. When she was in Primary 3 her parents divorced and her mother left home; her father then married another woman. She did not cope well without her mother, commenting: 'Our elder brothers took us up and raised us up and made decisions concerning our lives.' She then went to stay in Kiyingi village where her other brother lived. He took her to another school in Kiboga where she remained from Primary 4 until Primary 7: 'I got very poor grades – actually I failed and my brother did not feel the worth of paying my fees anymore.' After Primary 7 her brother suggested she learn tailoring, and introduced her to a woman in Kiboga town who was to train her. She did housework for the woman for 2 years, but the woman never paid her as she had promised. This woman also had a nephew 'who she wanted me to marry against my will'. A friend helped her escape and took her to Mukono to work in a house. The police got involved, and on the third day in her new job the brother came 'in anger' and took her back to where her mother stayed: 'My mother too was not happy with me.' She was later taken to her father's home, but she was uncomfortable with the stepmother, so after a week she returned to her mother who took her to a cooking and tailoring course in Nsambya. Then her mother fell sick, and her elder brother took her again to help him sell matooke at a stall on Kawempe town. At one point the brother disappeared, and she ran out of money: 'I decided to do paid casual work for neighbours. I worked as a hotel attendant, serving customers [their] food.' It was during this time that she 'got a man'. Her husband was a casual labourer at building sites. His maternal grandparent was staying in Kawempe and offered him land in Luweero, and this is how they came to Luweero.

Disruption of schooling was even more extreme for participants in contexts affected by civil war, whose education profiles, to say nothing of the education to work transition, were anything but linear. The experience of a 35-year-old woman from Awach, Uganda, makes the point:

> I started schooling [in] Primary 1 when people were not yet in the camp. In 2002 I moved to Gulu town and started schooling at Laliya Primary School. I schooled for 2 years in this school [Primary 3 and Primary 4]. In 2004, I was transferred to Aworanga Primary School, which is still in Gulu town. I was in this school for 2 years too [Primary 5 and Primary 6]. In 2005, I moved back to Awach Sub-County and joined Awach Mission Primary School [Primary 7]. In this same year, I dropped out of school because there was no money now for school fees.

In general, it was remarkable that schooling figured in any prominent way in the young people's livelihood narratives, given that they were constantly navigating different kinds of low-paid work, and responding to wider crises, whether provoked by deteriorating family circumstances or wider social upheavals.

The interviews did not specifically probe the costs of schooling, but alongside many who talked simply about school fees, other costs that were mentioned included uniforms, examination fees, charges for photocopies, books, food and transport costs. Some talked about having to 'dig' for the teacher, so schooling was provided in return for their unpaid work. Overall, attending school was evidently a significant financial burden. In many cases it was shouldered, at least in part, by the young people themselves, at the same time as they were contributing to family income, and sometimes also to the costs of schooling for their younger siblings. We return in the concluding section to the evident gap here between the young people's accounts of their experiences of education and work, and international policy commitments to the availability of 'free' or 'affordable' quality education through the life course, along with the assumed separation between school and work.

The gendered landscape of education and work

As shown above, females and males are involved in both education and work, but there are important gender differences framing their engagement, which impacts differentially on their education. Work itself is mostly mapped to sex, in a conflation of sex and gender, so that different forms of work are viewed as appropriate for female or male youth. Although some vocational work such as tailoring can be done by both sexes, young men are typically involved in activities such as farming, barbering (male hairdressing), carpentry, decorating, motorcycle taxi, tiling and welding; while young women work in farming, hairdressing, manicure/pedicure, cooking, trading and different forms of hospitality, including bartending. Going against expectations of what kinds of work are done by a woman or man could put one's gender into question. For example, in Côte d'Ivoire, preparing food is generally seen as women's work, to the extent that a young man who cooked and sold food to pay for his school fees was called by his sister a 'girl boy'. This was explained as meaning someone who 'is a boy, but not a boy. I could say he is a girl' (26-year-old male, Soubré).

While all of the respondents work in one way or another, males are more able than females to engage in different kinds of paid work outside the home. In contrast, many females are routinely involved in unpaid domestic labour, alongside helping on the family farm, as this Nigerian female recounted:

> When I came back from school, I cooked food, washed clothes and did other home chores. Also, during the weekend I went to live with my parents in the farm settlement every Friday to help them out with farm work and then return home on Sunday. There were not many challenges. [...] I would have loved to continue with my education, but I couldn't due to lack of money, that was why I stopped schooling.
>
> (30-year-old female, Umumbo, Nigeria)

This young woman's account of her day contrasts with that of a Nigerian man, who explained how he was constantly and proactively engaged in a succession of different forms of paid work, which he used to finance his education:

> While I was growing up as a young boy, I always followed my parents to farm. At the same time, I learnt electrical artisan work while growing up to support myself. I assisted my father on the fishing activities and fished for myself. I did all kinds of daily jobs, like field maintenance job. I did charcoal processing job for some months. [...] Whenever the charcoal season is gone, I always looked for other job, like the maintenance in the big farm and assisting the tractor operators.
>
> (25-year-old male, Oba Oke, Nigeria)

Working outside the home and indeed outside the community – doing artisan work such as carpentry, electrical work, masonry or decorating – is possible for young men without wider recriminations. For females, the risks are much greater, as codes of moral propriety frame the work and livelihoods that are open to them. Their futures are dominated by the expectation that they marry; this was openly discussed as a question of family honour and dishonour, particularly if a young woman becomes pregnant outside of marriage. At times, marriage is represented as something that is arranged for young females; in other instances, as something they actively engage with:

> I am the first child of my parents. They were not well-to-do, though they were farmers but could not get basic subsistence. We don't always have what to eat, despite that they tried to give us basic education. When suitors started coming for my hand in marriage I decided to marry since I was the first child so that I can help my parents in training my younger ones at least up to secondary school.
>
> (30-year-old female, Igbariam Eziafor, Nigeria)

A married woman will move to her husband's home, and is no longer the responsibility of her parents. With a few exceptions, while maintaining their respectability, this typically ended their education. While some young men were living independently of their parents, for females this was much less common and was often associated with stigmatized forms of work and social marginalization. For example, in Daloa, Côte d'Ivoire, a 30-year-old sex worker recounted how her parents and the father of her child had been killed in the civil war. Finding herself destitute and without education, she had taken up sex work as she had no other way of making a living. She spoke of the opprobrium she suffered as she walked down the street, and even her clients refused to look at her. Similarly, a young female in Nigeria who lived independently and worked in a bar described not being respected in society. She spoke of the kind of life that a woman had to avoid, in ways that illuminated the complex moral codes associated with the work and education of young women. She acknowledged both the social status attached to being literate and the shame of becoming pregnant outside marriage:

> I wouldn't want to be an illiterate. An illiterate is someone who doesn't go to school at all. There are other ways someone can become illiterate, like someone who gets pregnant while in school.
>
> (19-year-old female, Umumbo, Nigeria)

Although not explored further in the interview, the association of pregnancy with 'becoming' illiterate is indicative of the ways young women are forced out of schooling if they fall pregnant, as well as sometimes finding themselves rejected by their families. The ways female elders talk about the education of their daughters indicated how moral codes idealize the 'good' marriage, and indeed provide the justification for young women's education:

> For my daughters, my advice always for them is to be educated with a skill and be employed so that they can get a good husband. In my culture the pride and honour of every woman is her husband. So, marriage for my daughters is very important.
>
> (46-year-old female, Umumbo, Nigeria)

Other female elders in the same context echoed similar sentiments, adding that having a skilled job with a steady income would mean young women could 'have a voice in the home' (70-year-old female, Igbariam Eziafor, Nigeria).

To some extent, education is viewed as creating the potential for change in the world of work and in the home. There is some acknowledgement that women are now taking up professional positions (as doctors, lawyers, teachers) and marriage is taking place at a later age, as a result of education. As discussed in Chapter 9 (this volume), having a professional career or a wage job was often central to the imagined futures of young rural women (and men), as part of changing relations that were important for the next generation:

> I have experienced being a housewife and found out that is not good at all. [...] I have decided that all my children will be educated and acquire at least an NCE [Nigeria Certificate in Education for teaching]. Even if they don't want to become a teacher, they can get job in other areas. I will not like my daughters to become a housewife tomorrow like I was before finding something to do.
>
> (30-year-old female, Igbariam Eziafor, Nigeria)

At the same time, the gendered landscape of education and work in the rural economies of the research sites showed that completion of secondary school – and sometimes primary school – was out of reach for many if not most young women, in part because of the burden of unpaid labour and the moral judgements that framed the kinds of work they could do outside of the home. Many young men also struggled to complete primary and secondary school, but for different reasons.

Conclusions: The Inequalities of Schooling and Work

The social landscape of schooling and work in SSA suggests that education is valued by young women and young men. Nevertheless, it remains inaccessible to many, despite national and international policy commitments to the Sustainable Development Goals and 'affordable' quality education throughout the life course. In contrast to the idealized life trajectory that sees schooling leading to work (as assumed in HCT), these are simply not sequential or mutually exclusive. Many young people start to work before they start school, and then routinely combine multiple forms of work with attending school. This work is used to pay fees and other costs of school, as well as for young people to support their family. Clearly against many policy assumptions and strategies for intervention, leaving school is not a question of lack of aspiration; many young people see education as important in their imagined futures. Overwhelmingly, they foreground the cost of education as the reason for leaving school. In the rural economies of SSA, education still, at least in principle, offers the promise of social mobility, but remains out of reach for many.

However, the gendered landscape of education and work impacts differentially on female and male youth. Again, this is not a question of individual deficit. Deeply gendered expectations of what is appropriate for young women and men shape their trajectories through the intersecting spaces of home, school and work. In the process they sustain a patriarchal morality in which young women are overwhelmingly constructed as wives, mothers and homemakers, who further contribute in gendered and moralized ways to paid activities that are more routinely recognized as 'work'. Young women's education may be valued, but often in an instrumental way in terms of marriage capital; or the value of marriage may be considered in supporting the welfare of others. The importance of the family as a main source of support means that those who fall outside its norms are left vulnerable, forced into forms of work that bring them into disrepute and so compound their marginalization.

The narratives of rural young people raise many questions about the overwhelming reliance on HCT to understand the relationship between education and work. It does not account for social relations that fall outside its narrow economistic rationalities; these cannot address local social contexts of schooling and work, and the wider set of capitals that enable some to benefit from schooling, while excluding others; and it cannot attend to the intersections of schooling with different forms of work, including unpaid work and work done at school (e.g. 'digging' for the teacher). Young people's narratives also prompt questions about the school curriculum, what this proposes as ideals of life and work for young women and men, and how the rhythms of schooling accommodate rural livelihoods.

Importantly, contemporary concerns for 'quality' education – while clearly fully justifiable – seem far from addressing the matters highlighted in this chapter. A particular issue is the translation of the broader ambitions of SDG 4 into a much narrower set of targets. These misrecognize the politics of education and over-rely on learning metrics as a solution to complex social and cultural issues (Benavot and Smith, 2020). The narrow focus on pupil attainment also makes pupil assessment central to these metrics, and seems to assume it to be an unproblematic, predominantly technical issue. This ignores extensive critical commentary on educational assessment as a complex social practice. The danger here is that education is reduced to learning only what is easy to measure, in order to allow straightforward comparison of outcomes across substantially different contexts. The contemporary policy focus on learning also silences questions about what is to be learned, and by whom this is determined. As Biesta (2009, p.39) notes, the term *learning* 'denotes processes and activities but is open – if not empty – with regard to content and direction'. Significantly, such learning metrics also reduce gender to a female/male binary, in ways that yield no understanding of how gender is produced within different social and cultural contexts, whether in the family, work or schooling (Dunne, 2008).

Overall, the research illuminates the deep inequalities that structure the rural economy, and the part that school plays in sustaining them. To help address these, there is a clear need for in depth research into the intersections of work and schooling that does not assume schooling to be a black box, but explores its processes in a contingent and contextualized way, that is attentive to the dynamics of local social relations and their particular histories.

Notes

[i] Following ILO (2018b) the informal sector consists of 'units engaged in the production of goods or services, with the primary objective of generating employment and incomes to the persons concerned. [...] They are owned by individual household members or several members of the same or different households. Typically, they are operating at a low level of organization, on a small scale and with little or no division between labour and capital as factors of production.'

[ii] Those in 'vulnerable employment' include own-account workers and contributing family workers; the category is set against those who are in paid employment, who enjoy more job security and better working conditions (ILO, 2018a).

References

Adzahlie-Mensah, V. & Dunne, M. 2018. Continuing in the shadows of colonialism: the educational experiences of the African child in Ghana. *Perspectives in Education,* 36, 44–60.

Ansell, N. 2018. *Education Systems,Aspiration and Learning in Remote Rural Settings,* London, Brunel University.

Bakari, S. & Leach, F. 2007. Hijacking equal opportunity policies in a Nigerian College of Education: The micropolitics of gender. *Women's Studies International Forum,* 30, 85–96.

Ball, S. J. 1983. Imperialism, social control and the colonial curriculum in Africa. *Journal of Curriculum Studies,* 15, 237–263.

Becker, G. 1962. Investment in human capital: A theoretical analysis. *Journal of Political Economy,* 70, 9–49.

Benavot, A. & Smith, W. C. 2020. Reshaping quality and equity: Global learning metrics as a ready-made solution to a manufactured crisis. *In:* Wulff, A. (ed.) *Grading Goal Four: Tensions, Threats, and Opportunities in the Sustainable Development Goal on Quality Education,* Leiden, Netherlands, Brill/Sense.

Bennell, P. 1996. Rates of return to education: Does the conventional pattern prevail in sub-Saharan Africa? *World Development,* 24, 183–199.

Biesta, G. 2009. Good education in an age of measurement: on the need to reconnect with the question of purpose in education. *Educational Assessment, Evaluation and Accountability,* 21, 33–46.

Bolibaugh, J. 1972. *Educational Development in Guinea, Mali, Senegal, and Ivory Coast,* Washington, DC, Institute of International Studies.

Bourdieu, P. 1977. *Outline of a Theory of Practice,* Cambridge, UK, Cambridge University Press.

Bourdieu, P. & Wacquant, L. J. D. 1992. *An Invitation to Reflexive Sociology,* Cambridge, UK, Polity Press.

Boyden, J. 2013. 'We're not going to suffer like this in the mud': educational aspirations, social mobility and independent child migration among populations living in poverty. *Compare: A Journal of Comparative and International Education,* 43, 580–600.

Bryant, K. D. 2020. *Education and Politics in Colonial French West Africa. Oxford Research Encyclopedia,* Oxford, UK, Oxford University Press.

Butler, J. 1990. *Gender Trouble. Feminism and the Subversion of Identity,* London and New York, Routledge.

Butler, J. 1997. Merely cultural. *Social Text,* 52/53, 265–277.

Camfield, L. 2011. 'From school to adulthood'? Young people's pathways through schooling in urban Ethiopia. *European Journal of Development Research,* 23, 679–694.

Chuta, N. & Morrow, V. 2015. *Youth Trajectories through Work and Marriage in Rural Ethiopia. Working Paper 135,* Oxford, UK, Young Lives.

Clignet, R. P. & Foster, P. J. 1964. French and British colonial education in Africa. *Comparative Education Review,* 8, 191–198.

Connell, R. W. 2005. *Masculinities (2nd ed.).* Cambridge, UK, Polity Press.

Crossouard, B. & Dunne, M. 2020. Gender and education in postcolonial contexts. *In:* Mayo, C. & Mbisi, T. (eds.) *Oxford Research Encyclopedia of Education,* Oxford, UK, Oxford University Press.

Dunne, M. 2007. Gender, sexuality and schooling: Everyday life in junior secondary schools in Botswana and Ghana. *International Journal of Educational Development,* 27, 499–511.

Dunne, M. 2008. Gender docility: the power of technology and technologies of power for low-income countries. *In:* Chronaki, A. (ed.) *Mathematics Technologies Education. The Gender Perspective,* Volos, Greece, Thessaly University Press.

Dunne, M. & Adzahlie-Mensah, V. 2016. Subordinate subjects: The work of the hidden curriculum in post-colonial Ghana. *In:* Wyse, D., Hayward, L. & Pandya, J. (eds.) *The SAGE Handbook of Curriculum, Pedagogy and Assessment,* London, Sage.

Dunne, M. & Ananga, E. D. 2013. Dropping out: Identity conflict in and out of school in Ghana. *International Journal of Educational Development,* 33, 196–205.

Dunne, M., Leach, F., Chilisa, B., Maundeni, T., Tabulawa, R., Kutor, N., Forde, L. D. & Assamoah, A. 2005. *Gendered School Experiences: The Impact on Retention and Achievement in Botswana and Ghana,* London, DFID.

Dunne, M., Humphreys, S. & Leach, F. 2006. Gender violence in schools in the developing world. *Gender and Education,* 18, 75–98.

Farrugia, D. 2018. *Spaces of Youth: Work, Citizenship and Culture in a Global Context,* London, Routledge.

Feldman, H. 2016. The long shadows of Spanish and French colonial education. *Kyklos,* 69, 32–64.

Finlay, J.E., Efevbera, Y., Ndikubagenzi, J., Karra, M. & Canning, D. 2019. Reframing the measurement of women's work in the sub-Saharan African context. *Work, Employment and Society,* 33, 518–528.

Guidi, P. 2018. 'For good, God, and the Empire': French Franciscan sisters in Ethiopia 1896–1937. *History of Education,* 47, 384–398.

Healy, M. E. 2011. 'To Control Their Destiny': The politics of home and the feminisation of schooling in colonial Natal, 1885–1910. *Journal of Southern African Studies,* 37, 247–264.

Hossain, N. & Hickey, S. 2019. The problem of education quality in development countries. *In:* Hossain, N. & Hickey, S. (eds.) *The Politics of Education in Developing Countries: From Schooling to Learning,* Oxford, UK, Oxford University Press.

Humphreys, S., Moses, D., Kaibo, J. & Dunne, M. 2015. Counted in and being out: Fluctuations in primary school and classroom attendance in Northern Nigeria. *International Journal for Educational Development,* 44, 134–145.

ILO 2017. *World Employment Social Outlook Trends 2017,* Geneva, International Labour Organization (ILO).
ILO 2018a. *Paid Employment Versus Vulnerable Employment. A Brief Study of Employment Patterns by Status in Employment. ILOStat Spotlight on Work Statistics No. 3*, Geneva, International Labour Organization (ILO).
ILO 2018b. *Women and Men in the Informal Economy: A Statistical Picture (Third Edition),* Geneva, International Labour Organization (ILO).
Ilon, L. 1994. Structural adjustment and education: Adapting to a growing global market. *International Journal of Educational Development,* 14, 95–108.
Inoue, K., Di Gropello, E., Taylor, Y. S. & Gresham, J. 2015. *Out-of-School Youth in Sub-Saharan Africa. A Policy Perspective. Directions in Development,* Washington, DC, The World Bank.
Jomo, K. S. & Fine, B. (eds.) 2006. *The New Development Economics: After the Washington Consensus,* London, Zed Books.
Koolwal, G. 2019. *Improving the Measurement of Rural Women's Employment. Global Momentum and Survey Research Priorities. Policy Research Working Paper 8840,* Washington, DC, The World Bank.
Leach, F. 2008. African girls, nineteenth-century mission education and the patriarchal imperative. *Gender and Education,* 20, 335–347.
Maconachie, R. & Hilson, G. 2016. Re-thinking the child labor 'problem' in rural sub-Saharan Africa: The case of Sierra Leone's half shovels. *World Development,* 78, 136–147.
Marginson, S. 2019. Limitations of human capital theory. *Studies in Higher Education,* 44, 287–301.
McClintock, A. 1995. *Imperial Leather: Race, Gender, and Sexuality in the Colonial Contest,* London, Routledge.
McDowell, L. 2014. Gender, work, employment and society: feminist reflections on continuity and change. *Work, Employment and Society,* 28, 825–837.
McGrath, S., Ramsarup, P., Zeelen, J., Wedekind, V., Allais, S., Lotz-Sisitka, H., Monk, D., Openjuru, G. & Russon, J.-A. 2019. Vocational education and training for African development: a literature review. *Journal of Vocational Education & Training*, 1–23.
Mies, M. 2014. *Patriarchy and Accumulation on a World Scale: Women in the International Division of Labour (2nd ed.),* London, Zed Books.
Mirembe, R. & Davies, L. 2001. Is schooling a risk? Gender, power relations, and school culture in Uganda. *Gender and Education,* 13, 401–416.
Montenegro, C. & Patrinos, H. 2014. *Comparable Estimates of Returns to Schooling Around the World. Policy Research Working Paper 7020,* Washington, DC, The World Bank.
Nilsson, B. 2019. The school-to-work transition in developing countries. *Journal of Development Studies,* 55, 745–764.
Nwauwa, A. O. 2020. Western education and the rise of a new African elite in West Africa. *In:* Spear, T. (ed.) *Oxford Research Encyclopedia of African History,* Oxford, UK, Oxford University Press.
Pateman, C. 1988. *The Sexual Contract,* Cambridge, UK, Polity Press.
Pryor, J. & Ampiah, J. G. 2003. *Understandings of Education in an African Village: the Impact of Information and Communication Technologies,* London, DFID.
Rai, S. M., Brown, B. D. & Ruwanpura, K. N. 2019. SDG 8: Decent work and economic growth – A gendered analysis. *World Development,* 113, 368–380.
Resnik, J. 2006. International organizations, the 'Education-Economic Growth' black box, and the development of world education culture. *Comparative Education Review,* 50, 173–195.
Scanlon, D. 1964. *Education in Uganda,* Washington, DC, Office of Education, US Department of Health, Education and Welfare.
Schultz, T. W. 1961. Investment in human capital. *American Economic Review*, 51(1), 1–17.
Stockwell, A. J. 2012. Examinations and empire, the Cambridge Certificate in the colonies, 1857–1957. *In:* Mangan, J. (ed.) *Making Imperial Mentalities: Socialisation and British Imperialism (2nd ed.),* Manchester, UK, Manchester University Press.
Sumberg, J., Flynn, J., Mader, P., Mwaura, G., Oosterom, M., Sam-Kpakra, R. & Shittu, A. I. 2020. Formal-sector employment and Africa's youth employment crisis: irrelevance or policy priority? *Development Policy Review,* 34, 428–440.
The World Bank 2006. *World Development Report 2007: Development and the Next Generation,* Washington, DC, The World Bank.
The World Bank 2014. *Aspiration Traps: When Poverty Stifles Hope,* Washington, DC, The World Bank.

The World Bank 2018. *World Development Report 2018: Learning to Realize Education's Promise,* Washington, DC, The World Bank.

The World Bank 2020. *The Human Capital Index 2020 Update: Human Capital in the Time of COVID-19,* Washington, DC, The World Bank.

Tikly, L. 2019. Education for sustainable development in Africa: a critique of regional agendas. *Asia Pacific Education Review,* 20, 223–237.

UN WOMEN & ILO 2020. *Spotlight on SDG 8: The Impact of Marriage and Children on Labour Market Participation,* New York and Geneva, United Nations (UN) Women and International Labour Organization (ILO).

UNESCO 2015. *Global Monitoring Report Regional Overview: Sub-Saharan Africa,* Paris, UNESCO.

UNESCO 2018. *Global Education Monitoring Report 2019. Migration, Displacement and Education: Building Bridges, not Walls,* Paris, UNESCO.

Vavrus, F. 2002. Uncoupling the articulation between girls' education and tradition in Tanzania. *Gender and Education,* 14, 367–389.

Wulff, A. 2020. Introduction: Bringing out the tensions, challenges, and opportunities within Sustainable Development Goal 4. *In:* Wulff, A. (ed.) *Grading Goal Four – Tensions, Threats, and Opportunities in the Sustainable Development Goal on Quality Education*, Leiden, the Netherlands, Brill/Sense.

Yeboah, T., Sumberg, J., Flynn, J. & Anyidoho, N. A. 2017. Perspectives on desirable work: Findings from a Q study with students and parents in rural Ghana. *European Journal of Development Research,* 29, 423–440.

8 Are Rural Young People Stuck in Waithood?

Marjoke Oosterom
Institute of Development Studies, Brighton, UK

Introduction

This chapter revisits Alcinda Honwana's concept of 'waithood' (Honwana, 2012), which has become central to much youth research and policy discourse in sub-Saharan Africa (SSA), North Africa and the Middle East. In academic articles and policy debates, the concept is now widely used as shorthand to describe the situation in which youth are being held back and becoming frustrated by their inability to attain social adulthood. Despite waithood being critically interrogated in the research literature (e.g. Kovacheva *et al.*, 2018), its growing popularity in policy and public discourse has not been accompanied by critical debate about the value of the concept itself. Rather, the idea that youth are 'in waithood' is now largely taken for granted, and the meaning of waithood is often reduced to inactivity, despite Honwana's recognition that youth are 'creatively harnessing all the means at their disposal to manage their lives' (Honwana, 2012, p.20).

One implication of the popular discourse of 'youth in waiting', is that it locates the problem with young people themselves, rather than with social, economic and political structures and relations. Changing this means shifting the perspective from seeing youth as the 'problem', to recognizing that society is unable to come to terms with the fact that young people today are facing structural conditions that undermine their potential to progress in their lives. Honwana (2012, p.58) herself pointed to neoliberal economic policies as having created the structural conditions in which young people find it hard to get work, while she contends that other markers of social adulthood (e.g. marriage and living independently from parents) are dependent on having work.

This chapter interrogates the concept of waithood based on the findings of qualitative research in Uganda, Ethiopia and Nigeria. Specifically, the analysis is based on 158 life history interviews. The argument developed here is that emphasizing 'the inability to enter the labour market' (Honwana, 2012, p.19) as the key feature of waithood downplays the other features of social adulthood that young people are attaining. While Honwana has presented a nuanced perspective on the lives of youth in her book, her eventual focus on the labour market as the key to adulthood has removed much of the subtle distinction from the discussion. In policy discourse, waithood is now closely associated with youth unemployment and underemployment. Further, as commonly used, waithood also pays insufficient attention to gender dynamics and the ways in which many young women attain social adulthood without entering the labour market. Specifically, it overlooks the active negotiation by young people of various social relationships and expectations.

DOI: 10.1079/9781789245011.0008

The chapter proceeds by first discussing Honwana's conceptualization of waithood and linking it to other relevant scholarship. It then situates the concept within the literature on rural youth and aspirations. Based on empirical data gathered from young rural women and men in Nigeria, Ethiopia and Uganda, the meaning of farming and other economic activities in their lives, particularly in relation to social status, is presented. Other avenues for claim making on social recognition, status and respect are then analysed, with a focus on marriage, family life and active citizenship. Throughout the chapter the gendered nature of the process of becoming a social adult is emphasized.

Waithood: The Debate

Honwana's concept

The concept of waithood was first introduced by Singerman (2007). It then gained traction following the publication of the book *Generation in Waiting* (Dhillon and Yousef, 2009), which argued that institutions were failing young people in the Middle East and North Africa, leaving them 'in waiting'. However, it was Honwana's (2012) book *The Time of Youth* that brought waithood to centre stage and led to its integration into policy discourse. She explains the concept of waithood as:

> [...] a prolonged and uncertain stage between childhood and adulthood that is characterized by their inability to enter the labour market and attain the social markers of adulthood [p.19]. [...] Waithood is a neither-here-nor-there position in which young people are expected to be independent from their parents but are not yet recognized as social adults. No longer a brief transitional stage in the life-course, waithood is becoming a permanent condition, as many young people remain stuck in this in-between situation. Indeed, waithood is becoming a new but socially attenuated form of adulthood [p.20].
>
> (Honwana, 2012, pp.19–20)

In this conceptualization, the emphasis is on having 'not yet' reached social adulthood (as it is understood within a given society), while it is also suggested that labour market activity is the primary pathway through which it is reached. The 2007 World Development Report, *Development and the Next Generation*, conveys a similar message when it states that the period of youth is the key period for transition from puberty to economic independence (The World Bank, 2006, p.2). While the report acknowledges other transitions to adulthood (including learning, staying healthy, forming families and exercising citizenship), it clearly prioritizes the transition to productive citizens who contribute to economic development.

A key aspect of waithood is its relational dimensions: the connection between economic activity and achieving relative autonomy, and between aspirations and expectations in relation to economic activity and its outcomes. As put forward by Honwana, waithood implies that economic activity is what enables young people to live relatively independently from parents, get married and provide for a family – markers that are almost universally important for social adulthood (Durham, 2004). The ability to attain these markers earns a person status and recognition, which again underlines that waithood is a relational concept. Quoting young people from across the African continent, Honwana conveys the frustration and anxiety of young women and men who are unable to attain such relative autonomy and status (Honwana, 2012, pp.19–38). Leavy and Smith (2010) point out that expectations and aspirations in relation to the outcomes of economic activity are strongly influenced by socio-cultural factors. Both adults and youth want to live up to the ideal type of social adulthood. Young people have clear ideas about what constitutes 'respectable work', which means that if someone gains financial independence based on low-status economic activities, they may still lack the social status of a respectable adult. Of course, ideas about 'respectable work' can be expected to change over time. White (2019, p.15) underlines the importance of the relational dimension of youth aspirations: 'We cannot fully understand what young people want for themselves without understanding what they want for their families and communities.'

Langevang (2008, p.2044) examines perspectives on social adulthood among young Ghanaians and finds that in their view, youth is about 'becoming somebody' by maturing socially and economically, in order to get married and have a family. She also finds that young people adhere to notions of neat transitions from education to a formal job, even when these are highly unlikely. Many studies find that youth from poor and rural backgrounds aspire to formal employment in white- or blue-collar jobs (OECD, 2017; Yeboah *et al.*, 2017; Elias *et al.*, 2018). On the other hand, it is recognized that transitions experienced by most young people are fluid and complex, and these people often find themselves in multiple roles and transitions at the same time, moving in and out of work and education (Langevang, 2008; Locke and te Lintelo, 2012; Nilsson, 2019).

Numerous scholars, including Honwana herself, argue that neoliberalism has contributed to structural conditions in which young people find it hard to achieve their aspirations (Ansell, 2014; Bessant *et al.*, 2017; Sukarieh and Tannock, 2018). For example, Honwana states:

> The structural adjustment policies promoted by The World Bank and the International Monetary Fund in Africa failed to promote economic growth and job creation; [...]. While it is true that the quality of the educational system needs to be strengthened and should complement labour-market needs, the key issue is the lack of enough jobs.
>
> (Honwana, 2012, p.58)

While Honwana repeatedly refers to structural adjustment programmes of the past, other scholars have argued that neoliberal policies continue to inhibit job-rich economic growth, as such policies tend to stimulate profit generation over job creation (Bessant *et al.*, 2017; Sukarieh and Tannock, 2018). Mueller *et al.* (2019, p.10) conclude that: 'Africa does not necessarily face a youth challenge, but rather the broader challenge of promoting inclusive growth and decent employment in today's competitive global economy' (see also Betcherman and Khan, 2018; Fox *et al.*, 2020; Sumberg *et al.*, 2020). Presenting a broader critique of the debate on work, Ferguson and Li (2018) argue that because a majority of the world population is not in formal employment, a profound analytical 'decentring' of waged and salaried employment as the presumed norm is required. Policies focused on youth employment have, however, not focused on addressing these structural conditions (Sumberg *et al.*, 2020). Instead, interventions have prioritized training and skill development, and entrepreneurship, which keep youth 'busy' without creating new jobs (see also Fox and Kaul, 2017). A narrow conceptualization of waithood as 'youth in waiting' perpetuates the idea that interventions need to keep youth occupied, without altering the adverse structural conditions in which the chances of finding meaningful employment are very low, especially in rural areas.

Africa's rural population is expanding despite ongoing urbanization, and so is the rural workforce (Losch, 2016; Mueller *et al.*, 2019). Agriculture, and particularly small-scale farming, is and will remain the most important source of adult and youth employment, currently accounting for 44% of all employment in Africa (Filmer and Fox, 2014, p.29; ILO, 2017a). For rural youth, structural conditions like gerontocracy and patriarchy have been highlighted for their negative impacts on opportunities, for instance when land access is hampered by the older generation (Richards, 2005; Amanor, 2010; White, 2012; Andersson Djurfeldt *et al.*, 2019). It has also been noted that in some contexts, high population density resulting in pressure on land, inhibits the farming activities of young people (see Chapter 4, this volume; White, 2012). Further, their gendered positionality within the household influences how different young people engage in (family) farming, and the kinds of opportunities made available them (Andersson Djurfeldt *et al.*, 2019).

Agency, rural livelihoods and social markers of adulthood

What is problematic about the concept of waithood is the very word itself: it implies 'waiting' in the sense of inactivity. This is contradicted by numerous studies that show young people engaging in multiple economic activities, often simultaneously. The real 'waiting' may be for social recognition and respect, but in this sense, waiting is inseparable from 'growing up'. Indeed, Honwana (2012) herself does not suggest at any point that young people are doing nothing. The third chapter of *The Time of Youth*, 'Getting by', explores the wide range of activities in which they are involved, most of which are in the informal economy. She suggests that this demonstrates that young people are not giving up on their aspirations: '... [they] resourcefully take action in pursuit of a livelihood' (Honwana, 2012, p.61). It has been argued that aside from any direct financial reward, young people's social and economic actions allow them to make claims on family and other resources (Ferguson, 2015). Unfortunately, the distinction between waithood and inaction has been largely overlooked, and most policy and public discourse has appropriated the term in ways that suggests all youth are both inactive and stuck (ILO, 2017b).

Especially for urban youth, studies have demonstrated how they are getting by, while at the same time clearly conveying their feelings of disillusionment and frustration, and the sense that they want more out of life (Fuh, 2012). For young people living in rural areas, it is important to understand the meaning and sentiments associated with farming and other rural economic activities; and what kinds of social and economic outcomes they hope for. Sommers (2012) finds that young men in rural Rwanda are unable to attain the idealized marker of social adulthood: constructing a home to which they can bring a wife. Note that here, it is the lack of land on which to construct a house that is the problem, not a lack of paid work. Both the real and expected outcomes will be influenced by social relationships and will be context specific (Sumberg *et al.*, 2012). The conditions young people find themselves in influence their aspirations, which may, over time, be adjusted to align with what is feasible (Leavy and Hossain, 2014). This includes the opportunities and social expectations that come with their gendered position within the household and community (Andersson Djurfeldt *et al.*, 2019). While a common assumption is that young people are not interested in farming, existing studies highlight that many young people would farm if working conditions were better and financial returns were higher (White, 2012; Leavy and Hossain, 2014; Sumberg *et al.*, 2017; Elias *et al.*, 2018, p.90; White, 2019), or they already combine farming with other activities (see Chapter 3, this volume; Yeboah *et al.*, 2020). For rural youth in Ethiopia, Tadele and Gella (2012) find that the desirability of farming depends on a host of factors including whether someone is still in school or not, gender, age and level of formal education. Staying in Ethiopia, Bezu and Holden (2014) find that those young people whose parents have land are more likely to plan for a future in agriculture.

While young women and men in rural areas may have difficulty realizing their economic independence, many are attaining other markers that are part of the transition to adulthood: cohabiting, starting families, taking care of relatives, and assuming other responsibilities (Berckmoes and White, 2014). None the less, formalizing a marriage appears to be a general aspiration among young people because it generates social standing and respect (Berckmoes and White, 2014; Reynolds-Whyte and Acio, 2017). For young women, cohabiting might allow escape from a challenging home or family situation, but new vulnerabilities may arise if the relationship breaks down, especially if children are involved. While returning to the parental home may be an option for some, social institutions to support children of separated parents are generally lacking or inadequate (Berckmoes and White, 2014; Hopwood, 2015; Reynolds-Whyte and Acio, 2017).

Studies that have used the concept of waithood have emphasized young people's economic activities and social status (Langevang, 2008), while civic–political activities have received less attention. However, one important dimension of the transition to adulthood is becoming an active citizen, especially when the meaning of citizenship is extended beyond voting, to a much wider range of engagements in public life and decision making (Wood, 2014). It is clear that young people across Africa make their voices heard (Resnick, 2019). Honwana (2012) herself dedicates a full chapter to youth politics, focusing especially on urban protests, where young people appeared to be in the vanguard. Other studies have shown young people's active political engagement at the micro level. For example, Jeffrey and Dyson (2014, 2016) use the concept of 'prefigurative politics' to explain the actions of educated, unemployed and underemployed young males in India, who negotiate with local state authorities to advance the interests of their communities. In a study of student politics in India, Jeffrey and Young (2012) conclude that 'waiting' does not imply passiveness, as students belonging to different castes remain politically active while having no economic prospects.

Studies of youth citizenship in rural settings are limited, but existing evidence does suggest similarities to what has been observed in urban areas. Often, the focus is on civic activities that help improve the welfare of the community and strengthen local networks. Turner (2015) shows how young people in Supingstad, in South Africa's north-west province, actively negotiate the powers and authority of chiefs and mobilize against corruption. Korzenevica (2016) has documented how Nepalese youth periodically come home to take over the farm work, so that their parents can participate in local political meetings. As returning home reflects a conscious decision about the importance of public life, she argues it must be considered an act of citizenship.

What is revealed from the existing literature, is that discussions of waithood have been dominated by a focus on narrowly defined economic activities, without properly connecting them to social and civic–political lives. Honwana herself has emphasized young people's agency in making a life when she discusses their political and social actions, and claims to social status and citizenship. However, her emphasis on young people's inability to enter the labour market unduly limits the understanding of the multiple pathways that young people use to make claims on society in order to enhance their social status. This is not to underestimate the importance of work in many young people's lives, but the almost exclusive focus on work means other actions and claims to social recognition are overlooked or undervalued.

The remainder of this chapter takes a closer look at these other dimensions of young people's lives. The sections that follow present the findings of life history interviews with young women and men in rural areas of Uganda, Ethiopia and Nigeria. Field work was conducted in four different sites in each country (see Chapter 2, this volume). All interviews focused on the factors that led to the economic and other activities the interviewees were engaging in at the time of research in 2018 and early 2019. They were also asked about their imagined futures, and their life expectations. The interviews emphasized the role of family and relatives, as well as education and mobility. The findings show that young women and men are making claims to respect and social recognition in various ways, with engagement in economic activity being only one. Starting a family, taking care of elders and relatives, continuing in education, and being active in public life are also important avenues for gaining respect and status.

Too Busy to Wait

Work

This section first addresses farming activities and subsequently other economic activities in which rural young women and men are engaged.

Farming is part of the upbringing of most young people in rural SSA. From an early age, both girls and boys help on the farm, along with other domestic and economic activities before and after school, during weekends and holidays, and sometimes instead of going to school. The ways in which farming figures in imagined futures, depends on a number of factors. For many, there is no other option but to farm, as there are few other opportunities. Other important factors are whether farming and non-farm activities generate enough income to build a house, eat, pay for education, and 'progress in life'. Whether one can meet expectations in relation to social becoming – including the ability to pay for one's own education or for the education of one's children – through farming and livestock rearing is, for obvious reasons, related to questions of access to and quality of land and other resources, as well as access to markets.

In Ethiopia, there were differences between areas where agricultural activity was somewhat profitable and where it was not. In Wondo Genet, many young people were involved in producing and trading khat, on family land or through sharecropping. For many, especially if they had relatively large plots, this generated sufficient income to pay for their children's education and/or continue in education themselves. Those who were able to save money imagined themselves eventually building houses in town, while maintaining their links to rural areas. In Kuyu *woreda*, on the other hand, all young people referred to the shortage of land and the fact that much of the land was infertile: 'The main problem in our area is lack of land. Not having land is not having life,' said a 27-year-old woman. Land shortages and lack of fertile land clearly affect how young people imagine their prospects, given that opportunities for further study are limited. They rarely have enough good quality land through their families in this area, and sharecropping or other contractual arrangements are considered to be too disadvantageous. In fact, those who have no land or only limited land are longing for access to land. Narratives about urban migration and – for those who can afford it – migration to the Gulf states, are more common in this area.

It is clear that agriculture is considered hard work, while the conditions in which it is done are associated with low status and potential harm to one's health in the longer term. While many are proud when their harvests are good, they do not feel respected. In the Acholi region of northern Uganda, the phrase '*Atye ka yele*' [I am struggling] is used frequently and in various ways to describe rural lives: '*Atye ka yele kede pur*' [I am struggling with farming], '*Atye ka yele ki culu kwan pa lutino na*' [I am struggling to pay for education of my children], '*Atye ka yele ku biacara mamega*' [I am struggling with my business]. These phrases indicate the difficulties young people face, while they also show how hard they are trying. There is considerable uncertainty about whether farming can generate sufficient food and income for even the most basic subsistence in this poor region.

Young women and men in Wondo Genet *woreda* in Ethiopia are hoping to 'upgrade' their agricultural activities by introducing new technology. One young man explained how he previously did manual labour, but because his income increased, he could buy equipment:

> When you face a problem or have few choices, you do everything. I did manual labour as a last resort. Now I have changed it [I no longer do this] because it takes much energy and reduces one's lifespan. Now, I pump water through a generator to support my farming.

With money earned by using a horse and cart to transport people's goods, this same respondent funds his weekend college. He hopes to buy a motorcycle, with the hope that it will make everything he does less tiring.

In Nigeria, young women and men often say that they want to stop farming because the manual labour in the hot sun is too demanding, and even harmful. Others, however, talk of their love for farming and how they take pride in what they are able to produce. This is especially so when farming offers prospects of a respectable life, including producing enough to (eventually) marry and have a home. For example, one 26-year-old male from Idi Amu, Nigeria, who is the only one in his family to have gone to university, was farming while working as a teacher, and earning additional money as a photographer at parties. He said:

> I love farming, I cannot drop it for anything. I also believe that an average Nigerian is a farmer. When Obasanjo was the President he was farming and even till now. To be sincere with myself, I would like to do more teaching and farming. The reason I want to do it more is that I'm envisaging of becoming rich from the two.

Referring to both 'average Nigerians' and the President as farmers, he underlines the respectability of farming. While his parents and siblings also farm and do other things, he acknowledges that because of his degree he was expected to provide relatively more food and money to the family. He wants to honour these expectations and be a responsible man. He lives on his own, but his fiancée often stays with him, and he provides for her: 'It is my responsibility despite the fact that I have not even married her.' The story of this young man illustrates how he is crafting his identity as both a farmer and respectable member of the community.

Also in Nigeria, young women and men are hoping agriculture will become more mechanized. 'I cannot leave the farm work no matter how rich I am, even if I am extremely rich, I will go into mechanized farming,' said one man (25 years old, Idi Amu). When given the opportunity, many indicate they would like to go and work in commercial farming areas where more technology is used. Comparing his situation to the Ilorin area, where large-scale, mechanized farming is common, a young man from Oba Oke, says:

> The youth in Ilorin are more exposed and aware of mechanization for production and this beautifies agriculture. To me, manual farming is a barrier, because agriculture is more interesting to practice using mechanization. Youth in Ilorin who are around 20 years of age own 20 acres of farmland.
> (25-year-old male, Oba Oke, Nigeria)

The examples cited here demonstrate that the question of whether youth do or do not want to farm is far too simplistic. More often than not, farming is done under hard and challenging conditions. There is a strong sense that more technology would change this, and a clear desire for farming to be more respected. The future that young people imagine is one where farming allows them to marry, raise a family, put their children in good schools, and (for some) enable the construction of a home 'in town'. Many would want to remain involved in farming, but preferably supervising and managing farm labour rather than doing the work themselves.

In each of the three countries, a small percentage of the youth interviewed engage only in farming: the majority are engaged in multiple farming, off-farm and non-farm activities. Young people in each location list numerous activities, including wage work and self-employment, ranging from low-paid activities such as charcoal burning and brewing, to more rewarding pursuits like trading and transport. As with farming, certain non-farm activities are seasonal and cannot be relied upon for a secure and stable income. Opportunities for formal employment are rare and only a few of the young people work as teachers or nurses in local schools and facilities.

The observation that most young people farm one way or another, and that many also engage in one or more off-farm or non-farm income-generating activities, suggests that youth engagement with the labour market is pervasive (for more on the microdynamics of this engagement, see Carreras *et al.*, 2020). Yet, the returns from these activities are often low, and they offer no clear prospects for upward mobility. However, they do offer (sometimes long-term, sometimes temporary) opportunities for income generation, and might lead to activities that are better paid and of higher status.

The opportunities for engaging in off-farm and non-farm activities are gendered. In all four sites in Uganda activities such as cooking in canteens, brewing and market trading are considered more appropriate for women, while mechanics roles are considered more appropriate for men. In fact, the findings for Uganda showed that young women have fewer options when it comes to non-farm activities, as some roles are considered exclusive to young men. Findings suggested a kind of hierarchy in economic activities in terms of the social standing they could generate. As demonstrated for farming, young women and men tend to have clear ideas about non-farm activities, and to what extent these activities are 'decent' and what makes them respectable. Thus, working in a bar at night serving male customers, was 'indecent' for women in Uganda, but that did not necessarily stop them. For young men in Nigeria, engaging in trade enhanced their opportunities for mobility, which they associated with greater autonomy. However, even if the activity generated a substantial income, like in the case of certain motorcycle taxi riders, young men felt frowned upon by (adult) family members as this kind of work was not considered a proper 'adult' job and was not given much respect. Young women in both Nigeria and Uganda who engaged in tailoring reported they preferred the cleanliness of their work, as opposed to 'dirty farm work', and also the increased social standing associated with providing a service that required 'professional' interactions with customers.

While many young people are involved in more than one economic activity, we encountered numerous young people who blame the youth themselves for any lack of progress in their lives. For example, a young man from Nigeria with a BA degree said:

> It is very sad that young people are facing challenges today but part of our problem is laziness, as a result of what is happening in the society; in this era there is no youth that wants to do hard jobs when there are cunning ways to make quick money. Let us speak the truth, young people do not want to work but want to make big money by means of exploitation and other bad means.
>
> (26-year-old male, Idi Amu, Nigeria)

Thus, the notion of 'lazy youth' is not just a perception that adults have of youth, but it also circulates among the young people themselves. Better-off and better educated youth may be reluctant to acknowledge the structural barriers that make it harder for their impoverished or otherwise disadvantaged peers to progress. A struggle for recognition and respect is central to relationships across socio-economic classes, and not limited to youth.

Marriage and family life

While many adults opined that young people were no longer respecting certain 'traditional' values, it was clear from our research that most youth are working hard to earn respect and meet society's expectation of them. In all three countries, taking care of elders, especially parents, is important for young people. An Ethiopian woman from a relatively better-off family, which was supporting her in university from income generated by growing khat, said:

> [After university] I want to live with my family for some time because they are suffering with me at the present. I would like to dedicate 80% of my income to my parents. After that, I would pray to have a good husband. At that time, out of my salary, I would invest 70% in my own family and the remaining in my parents. I would never stop supporting my family.
> (18-year-old female, Wondo Genet, Ethiopia)

Similarly, many young Nigerians said how important it was to 'be a blessing to' and to 'honour' their parents. The possibility of moving elsewhere in search of work involved consideration of what it would mean for parents and other family members.

Supporting the family extends beyond parents and siblings. In the Acholi region of northern Uganda, which was affected by over 20 years of armed conflict, many young people, often while still very young themselves, were responsible for the upbringing of younger siblings and the children of relatives. As guardians they show real commitment in paying for the schooling of their own and foster children, which in this context can mean a significant sum of money despite the promise of Universal Primary Education that was supposed to cut the costs of primary education (Levine, 2016). Even if a foster child still has a parent, in many cases, their young carers report that they do not receive any kind of support from them. How heavy the responsibilities can become is illustrated by a 21-year-old female in Awach who has three children with her husband (6, 5 and 2 years old). Four years ago, they started caring for two of her late sister's children (12 and 8 years old), who had been abandoned by their father. Then 2 years ago, they also took in her brother's children (6 and 3 years old), after he moved to Kampala to look for work. He stopped sending money after some time and no longer even calls. All of the foster children are of school age and the income the family generate through farming and the bit of tailoring work the woman does is insufficient to cover the tuition and healthcare costs for everyone.

While many young people start families as cohabiting couples, formalizing a marriage remains something to which most youth aspire. However, it was also clear that many think very carefully about what is needed before getting married. Considerations may be tactical or strategic, and connected to other opportunities, or events. For example, an Igbo woman in Igbariam Eziafor, Nigeria, said she decided to marry when she was 23 years old, because she was the first born and wanted to help her younger siblings through school. She calculated that this would be possible because of the new economic activities that marriage was expected to open up. A young Nigerian man explained that he wanted to have a stable income before marrying:

> I have a cousin in Abuja who got married at the age of 27 and his income could not feed his wife, [...] she visited her parents for food and other things so her family could eat. [Therefore] marriage is not in my programme presently; such a situation [as my cousin's] would be a shame to me. As long as I cannot afford to feed additional people, I would not try marriage.
> (25-year-old male, Oba Oke, Nigeria)

In his view, marriage and economic progress need to go together, and he is not frustrated about not having yet married. At the same time, entering into marriage is not necessarily the result of being 'ready'. One Ethiopian woman said her mother married her off after the death of her father, and a young man explained that he decided to marry after his father died as he needed to maintain the labour capacity of the household. Also in Ethiopia, Wondo Genet, a 42-year-old woman spoke of how her father had tried to shield her from 'marriage through abduction' – when a young man 'steals' a young woman away from her family – by twice relocating her. He eventually decided to marry her off so he could at least choose her spouse.

Yet even in cases of marriage through abduction, some negotiation on the part of young women can be observed. For instance, a 20-year-old woman in rural Wondo Genet, Ethiopia, explained how she was abducted by her current husband when they were both in their mid-teens. Her father sued the man in court, saying it caused her to drop out of school, but she was adamant that the reason she stopped school in Grade 5 was her parent's lack of financial support:

> My husband was not punished; we reached reconciliation through elderly persons in the area. I just said 'I am now married and I cannot reverse it'. Since he said that he would support my education I did not say no to the marriage.

Since the husband had some barbering skills, both their parents supported the couple to start a barbershop. The husband then paid for the woman's barber training in the nearest large town. They run the barbershop together and managed to get involved in khat production through sharecropping using hired labour, which then enabled them to buy mobile phones in the city and resell them in their barber shop. She elaborates:

> By saving money from what we are getting now, we plan to buy land and build our own house, we think, in Hawasa town. If that is not possible, we might build it in Basha town. We want to expand all activities including the sale of mobile phones. He [her husband] says that he would continue his education, but he is more interested in his work. He wants to educate me and our child.
>
> (20-year-old female, Wondo Genet, Ethiopia)

Young women and men describe how education is more likely to be a factor to consider in their decisions about marriage than it was for their parents' generation. In the words of a Nigerian woman:

> When my parents were young, they were under their parents. Education was not much, if you manage to finish Grade 6, you had tried and you could stop there. But now, we want to further our education, we want to do this and that.
>
> (21-year-old female, Igbariam Eziafor, Nigeria)

A number of young women and men in Ethiopia who had relatively better incomes, for example two men from Wondo Genet, one 26 and the other 28 years old, reported wanting to postpone marriage and family life in order to pursue their education. A 28-year-old Nigerian man explained how he resisted pressure to get married even though his parents felt he already had sufficient income. He first wanted to pursue more education and expand his tailoring business and his farming. Across all research sites, young people valued education and linked attaining more education to enhanced social standing (see Chapter 7, this volume). Even if it did not lead to a 'proper' job, education helped them to gain respect.

The many examples of how young women negotiate marriage, other aspirations and social status show the gendered nature of the transition to adulthood. Almost all married women interviewed had children and were no longer 'youth' in socio-cultural terms, while they were still considered so in terms of chronological age. If they had married in their early teens, they had transitioned directly from childhood to adulthood. Overall, it was clear that young women have less influence than their male counterparts over when they marry, although the examples given above illustrate how some can nevertheless negotiate educational opportunities following marriage. In Nigeria, some young men who were interviewed were renting a room in town together while engaged in petty trading and running small-scale businesses. While this offered an 'in between' way of obtaining relative autonomy, it was clear they too desired formal marriage and their own homes. Young unmarried females generally had less room for manoeuvre to live independently from their families, except for moving away to pursue education. Parents preferred that their daughters live with older sisters or other relatives if they worked away from home.

While the vast majority of young women interviewed in all sites aspire to have a better income, either through farming or other activities, for them work is not their sole avenue to adulthood. Many feel they will reach adulthood once they are married and bear children. In other words, if they were 'waiting', it was more about waiting for marriage and/or childbirth, irrespective of the work they were doing. For men on the other hand, being an adequate provider was more closely connected to earning a good income and having respectable work. Their 'waiting' was about work that was sufficiently remunerative and respectable.

In Nigeria, several women reported that they had some degree of autonomy over their earnings, whether they were married or not, and this shaped their ideas about future possibilities for education and work. A single, 25-year-old Yoruba woman from Idi Amu was adamant: 'I will definitely be a supportive wife, but my husband will not be controlling my money.' In the same site, a married, 25-year-old woman with two children declared she controlled the earnings from her farming and *gari* processing:

> Though he [her husband] doesn't manage my money nor have power over my money, it is me who willingly supports him. And one thing I do is that before I use my money for anything, my husband must know about it.

Another 30-year-old female in Igbariam Eziafor, Nigeria, reported that at the age of 29 she successfully negotiated with her husband to start a National Certificate of Education (NCE) course.

Active citizenship

> Our government, both state and federal, are not concerned about the welfare of the youth, and that is why you still find very old men and women ruling in government.
>
> (23-year-old male, Igbariam Eziafor, Nigeria)

In many rural areas, young people are far away from centres of power, and even from 'local' authorities, and opportunities for interactions with state actors are limited. This does not mean young women and men are unaware of governance issues in their countries. In some places, rural youth are acutely aware of disparities and 'hierarchies in citizenship' that limit the opportunities available to them. This was observed in northern Uganda, where involvement of the national army in the region's armed conflict was seen as a deliberate attempt to undermine Acholi identity, as this ethnic group is linked with the political opposition. In Nigeria, citizenship status was reported as a factor that shaped opportunities. In Osun State for instance, some young people felt that being labelled as 'indigenous' or 'settler'[1] affected their farming opportunities, as it is easier for indigenes to access land. For instance, a young male indigene said:

> I have more opportunities as an indigene of this community than immigrants. We are the ones assisting them [incoming migrants]. Some immigrants have plans before migrating to this place, and we do not see most of them as farm labourers, except those that are good at harvesting palm kernels. Any other farming activities like cultivating and harvesting arable crops are mostly carried out by the indigenes of this community. Also, some of them are commercial transporters. We love most ethnic groups and we treat them like indigenes.
>
> (22-year-old male, Oba Oke, Nigeria)

Young settlers recounted how only forest land is made available to them, which takes time and money to clear before it can be cultivated. A 27-year-old man in Oba Oke reported that because of this, settler youth were 'discouraged', and he himself wanted to move to Benue State where he thought commercial farming would be both possible and profitable.

In Ethiopia, field research was conducted within a year of Abiy Ahmed becoming Prime Minister in April 2018, which marked a new era in Ethiopian politics. While some young people were unaware of any changes except in the leadership, many commented that the current situation was 'more free' and they could now speak more openly about political issues. They expressed hope that the country would now be peaceful and that the economic situation would improve. Among their expectations of the new government were improvement in services such as health and education, and job creation. Yet already within a year of the new administration being in place, frustrations were expressed about the lack of visible change in rural areas. In the words of one young woman:

> There isn't any change; nothing has changed in our life so far. We have not seen any change in the market under the *ganda* administration. [...] I want a better life. What else do we want? No, it [the administration] does not bring us land; it does not bring us salary. [...] It is true that for those who are educated, their salary may be raised. [...] But [...] land is already held by people; from whom will the government take [land] and to give it to those who do not have it?'
>
> (27-year-old female, Kuyu, Ethiopia)

A 30-year-old young man in Kuyu had completed Grade 9 and stopped because, according to him, members of the security forces made it difficult for students who were accused or suspected of political activism. He had lived in three different towns outside his own region to pursue paid work opportunities and engage in politics. For example, he had mobilized with others to protest against the government before Abiy Ahmed became Prime Minister, and he lost many friends who the government regarded as 'dissidents'. He had been imprisoned for

3 months for his activism, and during this time he started writing appeal applications for himself and others. He also worked to improve his writing skills. At the time of this research in 2018–2019 he was trying to settle down and take care of his three children by running a 'drinking house' offering drinks produced by his wife. He also occasionally earned money as a wage labourer at construction sites and had just started sharecropping. Meanwhile, he was earning some money for 'writing charges' for local court cases, motivated by his earlier human rights work: 'I serve my community through my writing skills while my community serves me by giving me an income.' Before the change in government, he challenged corruption in the *kebele*, for instance in its allocation of aid, land and support for farming, claiming he was often excluded from support as a result. Hoping Abiy Ahmed's government would lead to positive changes, he said:

> I expect much from this government. I hope this government will work for the people fairly, contrary to the previous government which used to trick people in many ways. I do not expect this government to provide me with food, but rather to facilitate conditions in which I can work, and what to use to work. Previously, I was excluded from political participation, but now I am selected as a committee member in the *kebele* for active political participation.
>
> (30-year-old male, Kuyu, Ethiopia)

With political activity being more concentrated in urban areas, few of our respondents were actively engaged in formal politics. There are also few channels through which to influence what matters in their lives, and if they do exist, they are dominated by the older generations. The institutions that govern access to land provide a powerful example, and in northern Uganda they are a real concern. Here, the first port of call for settling land disputes is customary leaders and elders, who provide few opportunities for young people (or adult women) to have a say. In turn, young people accuse them of corruption and nepotism. They spoke of how some land disputes are 'inherited' from their parents and other clan members, and reported that they are victims of threats, violence and theft as a result. These disputes sometimes have implications beyond the ability to access land. One young man in Awach told us how he was taken out of secondary school in town because his family wanted him to farm, specifically to prevent the land being claimed by another clan. While rural, more marginalized populations in general are far removed from loci of decision making, and poorly represented, for young people an additional hurdle is that their age further limits their opportunities to take part. This challenge is more profound for young women. There was a clear sense among some that they might never be in a position of influence. Besides the few who aspire to become 'politicians', the feeling is that rural people overall do not enjoy full and substantive citizenship.

Conclusions: Claim Making and Waithood Negotiation

Qualitative research from twelve sites across three countries in SSA demonstrate that waithood is not about waiting or lack of engagement with the labour market, but rather about hard work, negotiation and claim making. Young people actively negotiate different aspects of their lives, from various educational and work opportunities to marriage and other social relationships, social norms and expectations. Indeed, they are all looking for (better) economic opportunities that can increase their earnings. Yet they are also searching for more social recognition and respect for whatever they are doing, and actively seek to enhance their social standing by exercising agency in non-work domains, such as education, marriage and care of relatives.

When research directly engages with rural young people and their perspectives on work, it shows many do not reject rural livelihoods, but rather want opportunities that enhance the conditions in which farming and other rural work is carried out. It also shows that some negotiate their parents' expectations about when they should marry, which challenges the 'work in order to marry' understanding that is so central to the waithood argument.

It must be reiterated that Honwana's discussion of waithood is more nuanced than the claim that all youth are stuck: she acknowledges and elaborates the multiple economic and civic–political activities in which young people are involved. Moreover, she points at the structural conditions that adversely affect the opportunities available to young people. However, the emphasis on young people's inability to enter the labour market and the way the concept of waithood has been appropriated by policy actors has produced a narrative on youth and work that is far too narrow and does not reflect the realities of rural SSA. Policy approaches continue to be dominated by skill building and entrepreneurship programmes that cannot change the adverse structural conditions in which rural young people find themselves.

Note

[i] In Nigeria the term 'indigene' is used to refer to individuals who are considered to be the 'first' or indigenous inhabitants of a particular place, while 'settlers' migrated into the area at a later point in time. In many states of Nigeria, these labels are contested. The Constitution of Nigeria distinguishes between settler and indigene groups, though ambiguities remain and have resulted in contestations over rights, dependent upon the settler and indigene status, thereby leading to conflict (see Akanji, 2011).

References

Akanji, O. O. 2011. 'The problem of belonging': the identity question and the dilemma of nation-building in Nigeria. *African Identities,* 9, 117–132.

Amanor, K. S. 2010. Family values, land sales and agricultural commodification in South-Eastern Ghana. *Africa: Journal of the International African Institute,* 80, 104–125.

Andersson Djurfeldt, A., Kalindi, A., Lindsjö, K. & Wamulume, M. 2019. Yearning to farm – Youth, agricultural intensification and land in Mkushi, Zambia. *Journal of Rural Studies,* 71, 85–93.

Ansell, N. 2014. Generationing development. *European Journal of Development Research,* 26, 283–291.

Berckmoes, L. & White, B. 2014. Youth, farming and precarity in rural Burundi. *European Journal of Development Research,* 26, 190–203.

Bessant, J., Farthing, R. & Watt, R. 2017. *The Precarious Generation: A Political Economy of Young People,* London, Routledge.

Betcherman, G. & Khan, T. 2018. Jobs for Africa's expanding youth cohort: a stocktaking of employment prospects and policy interventions. *IZA Journal of Development and Migration,* 8, 13.

Bezu, S. & Holden, S. 2014. Are rural youth in Ethiopia abandoning agriculture? *World Development,* 64, 259–272.

Carreras, M., Sumberg, J. & Saha, A. 2020. Work and rural livelihoods: the micro dynamics of Africa's 'youth employment crisis'. *European Journal of Development Research* (doi.org/10.1057/s41287-020-00310-y).

Dhillon, N. & Yousef, T. M. (eds.) 2009. *Generation in Waiting: The Unfulfilled Promise of Young People in the Middle East,* Washington, DC, The Brookings Institution.

Durham, D. 2004. Disappearing youth: youth as a social shifter in Botswana. *American Ethnologist,* 31, 589–605.

Elias, M., Mudege, N., Lopez, D. E., Najjar, D., Kandiwa, V., Luis, J., Yila, J., Tegbaru, A., Ibrahim, G., Badstue, L., Njuguna-Mungai, E. & Bentaibi, A. 2018. Gendered aspirations and occupational trajectories among rural youth: A cross-regional perspective. *Journal of Gender, Agriculture and Food Security,* 3, 82–107.

Ferguson, J. 2015. *Give a Man a Fish: Reflections on the New Politics of Distribution,* Durham and London, Duke University Press.

Ferguson, J. & Li, T. M. 2018. *Beyond the 'Proper Job': Political-economic Analysis after the Century of Labouring Man', Working Paper 51,* Cape Town, PLAAS, University of the Western Cape.

Filmer, D. & Fox, L. 2014. *Youth Employment in Sub-Saharan Africa,* Washington, DC, Agence Française de Développement and The World Bank.

Fox, L. & Kaul, U. 2017. *The Evidence Is In: How Should Youth Employment Programs in Low-Income Countries Be Designed?,* Washington, DC, USAID.

Fox, L., Mader, P., Sumberg, J., Flynn, J. & Oosterom, M. 2020. *Africa's 'Youth Employment' Crisis is Actually a 'Missing Jobs' Crisis. Booke Shearer Series Number 9,* Washington, DC, The Brookings Institution.

Fuh, D. 2012. The prestige economy: veteran clubs and youngmen's competition in Bamenda, Cameroon. *Urban Forum,* 23, 501–526.

Honwana, A. 2012. *The Time of Youth: Work, Social Change, and Politics in Africa,* Boulder, CO, Kumarian Press.

Hopwood, J. 2015. Women's land claims in the Acholi region of Northern Uganda: What can be learned from what is contested. *International Journal on Minority and Group Rights,* 22, 387–409.

ILO 2017a. *Global Employment Trends for Youth 2017: Paths to a Better Working Future,* Geneva, International Labour Organization (ILO).

ILO 2017b. *A Robust Regional Economy for Improved Labour Market Outcomes,* Geneva, International Labour Organization (ILO).

Jeffrey, C. & Dyson, J. 2014. 'I serve therefore I am': Youth and generative politics in India. *Comparative Studies in Society and History,* 56, 967–994.

Jeffrey, C. & Dyson, J. 2016. Now: prefigurative politics through a north Indian lens. *Economy and Society,* 45, 77–100.

Jeffrey, C. & Young, S. 2012. Waiting for change: youth, caste and politics in India. *Economy and Society,* 41, 638–661.

Korzenevica, M. 2016. Young people navigating political engagement through post-war instability and mobility: A case from rural Nepal. *Geoforum,* 74, 19–28.

Kovacheva, S., Kabaivanov, S. & Roberts, K. 2018. Interrogating waithood: family and housing life stage transitions among young adults in North-West Africa countries. *International Journal of Adolescence and Youth*, 23, 441–456.

Langevang, T. 2008. 'We are managing!' uncertain paths to respectable adulthoods in Accra, Ghana. *Geoforum*, 29, 2039–2047.

Leavy, J. & Hossain, N. 2014. *Who Wants to Farm? Youth Aspirations, Opportunities and Rising Food Prices. IDS Working Paper 439*, Brighton, UK, Institute of Development Studies (IDS).

Leavy, J. & Smith, S. 2010. *Future Farmers: Youth Aspirations, Expectations and Life Choices. FAC Discussion Paper 013*, Brighton, UK, Future Agricultures Consortium.

Levine, S. 2016. *Livelihood Recovery in Post-conflict Northern Uganda. Working Paper 42*, London, Secure Livelihoods Research Consortium.

Locke, C. & te Lintelo, D. 2012. Young Zambians 'waiting' for opportunities and 'working towards' living well: lifecourse and aspiration in youth transitions. *Journal of International Development*, 24, 777–794.

Losch, B. 2016. *Structural Transformation to Boost Youth Labour Demand in Sub-Saharan Africa: The Role of Agriculture, Rural Areas and Territorial Development. Employment Working Paper No. 204*, Geneva, International Labour Organization (ILO).

Mueller, V., Thurlow, J., Rosenbach, G. & Masias, I. 2019. Africa's rural youth in the global context. *In:* Mueller, V. & Thurlow, J. (eds.) *Youth and Jobs in Rural Africa: Beyond Stylized Facts,* Oxford, UK, Oxford University Press.

Nilsson, B. 2019. The school-to-work transition in developing countries. *Journal of Development Studies,* 55, 745–764.

OECD 2017. *Youth Aspirations and the Reality of Jobs in Developing Countries: Mind the Gap*, Paris, Development Centre Studies, OECD Publishing.

Resnick, D. (2019) 'Troublemakers, Bystanders, and Pathbreakers. The Political Participation of African Youth'. In: *Youth and Jobs in Rural Africa: Beyond Stylized Facts*. Mueller, V. and Thurlow, J. (Eds.). Chapter 4, pp. 75–108. New York, NY: International Food Policy Research Institute (IFPRI) and Oxford University Press.

Reynolds-Whyte, S. & Acio, E. 2017. Generations and access to land in postconflict northern Uganda: 'Youth have no voice in land matters'. *African Studies Review*, 60, 17–36.

Richards, P. 2005. To fight or to farm? Agrarian dimensions of the Mano River conflicts (Liberia and Sierra Leone). *African Affairs*, 104, 571–590.

Singerman, D. 2007. *The Economic Imperatives of Marriage: Emerging Practices and Identities Among Youth in the Middle East. Working Paper 6*, Washington, DC and Dubai, Wolfensohn Centre for Development and Dubai School of Government.

Sommers, M. 2012. *Stuck: Rwandan Youth and the Struggle for Adulthood*, Atlanta, GA, University of Georgia Press, in association with U.S. Institute of Peace Press.

Sukarieh, M. & Tannock, S. 2018. The global securitisation of youth. *Third World Quarterly*, 39, 854–870.

Sumberg, J., Anyidoho, N. A., Leavy, J., Te Lintelo, D. J. H. & Wellard, K. 2012. Introduction: the young people and agriculture 'problem' in Africa. *IDS Bulletin*, 43, 1–8.

Sumberg, J., Yeboah, T., Flynn, J. & Anyidoho, N. A. 2017. Young people's perspectives on farming in Ghana: a Q study. *Food Security*, 9, 151–161.

Sumberg, J., Flynn, J., Fox, L., Mader, P. & Oosterom, M. 2021. Africa's 'youth employment' crisis is actually a 'missing jobs' crisis. *Development Policy Review* (doi.org/10.1111/dpr.12528).

Tadele, G. & Gella, A. A. 2012. 'A last resort and often not an option at all': Farming and young people in Ethiopia. *IDS Bulletin*, 43, 33–34.

The World Bank 2006. *World Development Report 2007: Development and the Next Generation*, Washington, DC, The World Bank.

Turner, R. L. 2015. Youth, 'tradition' and peace building: mobilising for just governance in rural South Africa. *Peacebuilding*, 3, 126–140.

White, B. 2012. Agriculture and the generation problem: rural youth, employment and the future of farming. *IDS Bulletin*, 43, 9–19.

White, B. 2019. *Rural Youth, Today and Tomorrow. IFAD Research Series 48*, Rome, International Fund for Agricultural Development (IFAD).

Wood, B. 2014. Researching the everyday: Young people's experiences and expressions of citizenship. *International Journal of Qualitative Studies in Education*, 27, 214–232.

Yeboah, T., Sumberg, J., Flynn, J. & Anyidoho, N. A. 2017. Perspectives on desirable work: Findings from a Q study with students and parents in rural Ghana. *European Journal of Development Research*, 29, 423–440.

Yeboah, T., Chigumira, E., John, I., Anyidoho, N. A., Manyong, V., Flynn, J. & Sumberg, J. 2020. Hard work and hazard: Young people and agricultural commercialisation in Africa. *Journal of Rural Studies*, 76, 142–151.

9 Young People's Imagined Futures

Thomas Yeboah[1], Barbara Crossouard[2] and Justin Flynn[3]
[1]*Bureau of Integrated Rural Development (BIRD), Kwame Nkrumah University of Science and Technology (KNUST), Kumasi, Ghana;* [2]*University of Sussex, Brighton, UK;* [3]*Institite of Development Studies, Brighton, UK*

Introduction

This chapter explores the futures that rural young people imagine for themselves, and how they relate to both their current engagement with the rural economy, and the narrative that suggests a widely held desire to abandon agriculture and rural areas. Commenting on this narrative, White (2012) suggested that any reluctance among young people was likely due to the unattractiveness of the agricultural sector in its 'current state'. He suggested various reasons why young Africans may be turning their backs on agriculture, including: 'deskilling' and the low status of farming and rural life; poor rural infrastructure and lack of government investment in the sector; and the (increasing) difficulty young people have in accessing land. Since this publication, other authors have contributed further ideas – and some data – as to whether and why young people may be turning away from agriculture and rurality (see also Chapter 3, this volume).

In addition to hard work, low status, limited returns and lack of land, some authors have pointed to higher level concerns. For example, Moore (2015) reports that young people may be concerned that farming offers only limited opportunity to contribute to their community or country. This is in line with findings by Yeboah *et al.* (2017), that secondary students in Ghana had a strong 'social ethos', with jobs being more desirable if through them young people could 'make the world a better place'. For Gastineau and Golaz (2016), young people in Africa find rural areas have little appeal. They report that there is a shift or a discrepancy between generations, with young people being more educated, more mobile, and more open to the world (e.g. through the propagation of information and communications technologies, ICTs), and aspiring to a different type of life than that of their parents. This links to Leavy and Hossain's (2014) notion of a 'generational break' in aspirations, largely due to the difference in educational levels.

On the other hand, research over the past decade has provided a somewhat more nuanced picture of youth aspirations, where young people, depending on the context, may have a more neutral, or even quite positive view of agriculture and rural life. For example, in Burundi, a country where 95% of the rural population is involved in agriculture, farming is still seen as desirable way of life, and as one of the few realistic livelihood options (Berckmoes and White, 2014). Despite the challenge of ever decreasing plot sizes, young people still aspire to farming futures, but hopefully under better conditions. In Zambia, young people carefully considered the pros and cons of agriculture and rural life, and still wished to engage in agriculture, especially if

 Youth and the Rural Economy in Africa: Hard Work and Hazard (ed. J. Sumberg)
DOI: 10.1079/9781789245011.0009

they could use animals for draught power, use more fertilizer and have access to electricity (Daum, 2019). Also working in Zambia, Andersson Djurfeldt *et al.* (2019) report that rural youth generally want to farm, particularly in a more modern and professional manner based on technical knowledge and skills. Kristensen and Birch-Thomsen (2013) report similar conditional aspirations. However, making this a reality hinges on young people's ability to access land; where this is not possible, young women and men face difficulties in gaining independence, and so seek off-farm opportunities. In a sense, these findings link to a recent analysis of data from the School-to-Work-Transition Surveys (SWTS) of the International Labour Organization (ILO), which included students (aged 15–29) from 11 African countries (OECD, 2017). The main takeaway from this study is that the students' views about their futures are linked to their social standing in society: less-well educated young men generally aspire to lower skilled jobs while those with more education aspire to more highly skilled jobs. It is notable that younger students also generally aspire to high-level jobs. Finally, Sumberg *et al.* (2017) used Q Methodology to explore two questions with rural secondary students in Ghana: 'What explains young people's attitude toward farming?' and 'What should be done about young people and farming?' Different perspectives are evident, focusing variously on the young people themselves and their desire for modern jobs; society's lack of respect for farmers; the need to modernize farming; and the lack of services and facilities in rural areas. The key conclusion of this study is that across both genders, young people have a largely negative attitude toward farming; but that this negativity is rooted in quite different perceptions and understandings. Some of these perspectives may be amenable to interventions such as awareness raising and training, others not. This suggests the need for a more targeted approach to policy and programmes that seek to address youth employment through the agricultural sector.

In this chapter, we seek to bring greater understanding of young rural Africans' perspectives on their futures. We draw on two complementary sources of data. The first is a study carried out under the Agricultural Policy Research in Africa (APRA) programme[i] in three sites (one each in Ghana, Zimbabwe and Tanzania) where commercialized agriculture is relatively well developed (Yeboah *et al.*, 2020). The second is the work funded by the International Fund for Agricultural Development (IFAD) in 16 sites across Uganda, Ethiopia, Nigeria and Côte d'Ivoire (see Chapter 2, this volume). Both studies focused on current engagement with the rural economy, how young people imagined their futures, and what they think might stop them from realizing the futures they imagine for themselves. We are particularly interested in whether young women's and men's imagined futures align with the orthodoxy suggesting that young people are not interested in agriculture and rural life. Beyond imagined future economic activities, including farm and non-farm, we explore their imaginings in relation education and migration. Rather than using the term 'aspirations', which can imply a desire for the future that is not necessarily situated within a social context, in the remainder of this chapter we use the concept 'imagined futures'. Essentially, in various ways, young people were asked: 'How do you imagine your future'?

The chapter is structured as follows. The next section provides a more detailed exploration of the notion of imagined futures. Methods are then discussed, followed by findings. The last section is a discussion and conclusion including key implications for policy and research.

Framing Young People's Imagined Futures

Much has been written about the aspirations of young people in SSA (Leavy and Smith, 2010; Anyidoho *et al.*, 2012; Locke and te Lintelo, 2012; Petesch and Rodríguez Caillava, 2012; Bernard and Seyoum Taffesse, 2014; Kosec *et al.*, 2014; Leavy and Hossain, 2014; Favara, 2017; Mausch *et al.*, 2018; Verkaart *et al.*, 2018). The focus on aspirations undoubtedly captures some aspects of young people's dreams, musings and hopes about their future. However, the concept of aspirations has also been critiqued for being vague, lacking conceptual precision, and failing 'to draw a compelling link between future goals or outcomes and young people's agentive efforts to work towards them' (Hardgrove *et al.*, 2015, p.163).

The focus on aspirations, and especially the suggestion of a 'poverty of aspiration' has also been critiqued as putting the blame on young people and ignoring or downplaying structural constraints (McKendrick, 2017; White, 2019). Work that has explored the possibility of affecting aspirations through policy (Kosec *et al.*, 2014) also reflects a similar instrumental approach meant to address individuals' aspiration deficits.

It has been argued that the notions of imagined futures and imagined selves offer valuable alternatives to aspirations (Markus and Nurius, 1986; Hardgrove *et al.*, 2015). As a very simple example, imagine two female students in a rural secondary school. The first is asked what she aspires to, and she responds that she wants to be a lawyer, end of story. The second is asked how she imagines her future, and she responds that she imagines herself as a lawyer, working alongside her auntie, who is herself a successful lawyer and who has offered to help her to enter the profession. Importantly, rather than the more individualized concepts of 'aspirations' and 'motivation', imagined futures are typically theorized in a relational way that understands youth agency and intentionality to be socially located, and to be shaped both by past experiences and contemporary circumstances. In other words, such imagined futures are themselves the products of young people's 'perceived, structured positions in society' (Hardgrove *et al.*, 2015, p.164).

Apart from illuminating the identity positions that youth would like to embrace as they grow older and the social structures that frame these, youth narratives about their imagined futures also provide considerable insights into 'the internalized socio-cultural values' that young people hold (Hardgrove *et al.*, 2015, p.164). In Bourdieusian terms, this implies that a young person's articulation of an imagined future reflects the value systems that permeate a particular field and provides insights both to the habitus of the individual – i.e. the social process through which individuals become socialized (Bourdieu, 1984) – and the different capitals (economic, cultural, social and symbolic) that are accorded significance within their particular context (Bourdieu and Wacquant, 1992). As a 'structure of preference', an imagined future can be considered a product of the 'individual collective history of agents' and their positioning within a particular field (Bourdieu and Wacquant, 1992, p.123). In other words, an imagined future transcends a naïve separation of structure and agency, which is often implicit in the discussion of aspirations.

From these perspectives, questions about how realistic or realizable an imagined future might be, or about the difference between aspirations and expectations, are not particularly relevant. Such questions are indicative of the tendency in youth studies to treat social constructions, that should be understood in relational terms, in overly realist or substantialist ways (Bessant *et al.*, 2020). As Ball *et al.* (1999) note, an imagined future can be 'relatively clear, relatively stable and relatively possible' or 'vague, relatively unstable and beset with uncertainties'. Importantly, rather than questioning the credibility of their participants' imagined futures, analysis by Ball *et al.* (1999) stresses the importance of family capitals in making some imagined futures sound realizable, in contrast to others that remain vague and poorly articulated.

It is the social and cultural capitals that become ingrained through exposure to a more or less privileged background, that allow some participants to talk with ease about the particularities of their imagined futures, whereas others have little practical awareness of the field so cannot show mastery of it. In Bourdieu's terms, individuals in this latter group lack a 'feel for the game'. Here again, we see the relevance of Bourdieu's understanding of the mutually constitutive relations of habitus, field and capitals in the production of the social, as opposed to more individualistic understandings of the self (Bourdieu, 1984).

Research drawing on imagined futures has been influential within education and studies focusing on the transition from education to employment (Ball *et al.*, 1999; Hardgrove *et al.*, 2015; Stahl *et al.*, 2020). As Stahl *et al.* (2020) state, the concept lends itself to an analysis of the gendered construction of possible selves, and what forms of femininity and masculinity are privileged in the overlapping 'regimes of value' (Skeggs, 2011) that are obtained within the social landscapes framing youth livelihoods. Based on such understandings, a central part of the research in both the studies reported on here was to seek out how rural youth imagined their futures, with respect to their lives in general, their work and their education.

Methods

This chapter draws on two complementary sources of data focusing on how youth engage in the rural economy of SSA. First is the APRA Youth Policy Study carried out in three commercialized rural agricultural sites in three country contexts: Ghana (Techiman North District, Brong Ahafo Region), Zimbabwe (Mazowe District, Mashonaland Central Provence) and Tanzania (Kilosa District, Morogoro Region). The focus of the study was on the steps and pathways through which young people came to engage with the commercialized rural economy and the outcomes associated with their efforts. Data were collected through individual interviews, with 35, 42 and 40 interviews being completed in Ghana, Tanzania and Zimbabwe, respectively. The same interview schedule was used across the three sites covering: (i) the background of the interviewee; (ii) a history of her/his economic activities; and (iii) plans for the immediate and more distant future (see Yeboah *et al.*, 2020).

Second is the IFAD Youth Research project. As detailed in Chapter 2 (this volume), this study included field work in 16 sites (four sites per country) in Uganda, Ethiopia, Nigeria and Côte d'Ivoire. The sample included young women and men with different educational experiences. Data were collected through life history interviews, livelihood interviews and photo–voice interviews with approximately equal numbers of female and male youth. All three types of interview asked about how the interviewee imagined her/his future. There were also a number of sex-segregated focus group discussions (FGDs) that explored youth livelihoods, the participants' imagined futures and how both were gendered, and the significance of education to young people. A number of adults were also interviewed.

The Imagined Futures of Rural Young People

The futures that rural young people at sites in Uganda, Ethiopia, Nigeria, Côte d'Ivoire, Ghana, Zimbabwe and Tanzania imagined for themselves centred around: (i) expanding and/or diversifying current economic activities; (ii) accumulating wealth or assets; and (iii) acquiring further education and obtaining a professional or salaried job.

Expanding and/or diversifying current economic activities

As highlighted in Chapter 3 (this volume) the majority of young people are involved in farming and/or livestock production in one way or another, often combined with other non-farm income-generating activities. Across all countries and sites, a central feature of the futures that many young people imagined for themselves was the expansion and/or diversification of current economic activities.

For most young people in Uganda, and for many in the other sites, this meant that agriculture was important in their imagined futures (Table 9.1 and Appendix Tables 9A–9D). Overall, agriculture was more likely to be important in the imagined futures of men rather than women (in most contexts, women were more likely to give importance to non-farming activities), and individuals with less education. In Nigeria, agriculture was more important in the imagined futures of young people in 'cold spots' compared to those in 'hot spots', i.e. areas of relative market disengagement and stagnation versus areas of more commercial activity and dynamism (Chapter 2, this volume, lays out the definitional criteria for these areas).

Some individuals foresaw increasing the scale of production, while others imagined diversifying into new, more lucrative crops, and a few imagined specializing, for example by establishing plantations of cashew, kola or oranges. For many, these plans for expansion and diversification are unexceptional: a modest increase in the area of land farmed or the number of head of livestock kept, or for example, starting out in poultry production. Others imagine a more expansive engagement with agriculture:

> I want to be a great farmer, cultivating rice and cowpea. [...] I will increase my scale of operation in the next 5 years. I will enter into commercial farming, by that time I will be matured financially. I will employ people to manage my farm, I will only come around to supervise the farm.
>
> (29-year-old female, Umumbo, Nigeria)

Table 9.1. Importance of agriculture in young people's imagined futures.

Country	Site	Agriculture important in imagined future (%)
Uganda	Awach [relatively Hot]	92
	Luweero [relatively Hot]	78
	Mucwini [relatively cold]	88
	Butuntumula [relatively cold]	65
Ethiopia	Wondo Genet [relatively hot]	35
	Bora [relatively hot]	71
	Jabi Tehnan [relatively cold]	52
	Kuyu [relatively cold]	38
Nigeria	Oba Oke [relatively hot]	41
	Igbariam Eziafor [relatively hot]	38
	Idi Amu [relatively cold]	69
	Umumbo [relatively cold]	59
Côte d'Ivoire	Soubré [relatively hot]	38
	Odienné [relatively hot]	44
	Daloa [relatively cold]	27

> I like every part of farming, cultivating maize, and *teff*. I planned to leave farming in the past but now it is benefiting so that I will do more in farming. I want to focus on cultivating *teff* and maize by renting-in the land and using selected seeds and chemical fertilizers. Farming is profitable and I think I will bring economic change. [...] Yes it [farming] is sufficient. We don't buy food from the market. Now, my income will be rising fast. Particularly, if I cultivate using the application of chemical fertilizers and selected seeds, my income from farming will be increasing and sufficient to feed my family. Thus, I hope that there will be good things in farming. I can feed my family well and produce surplus for the market even.
>
> (25-year-old male, Kuyu, Ethiopia)

Some young people frame their future interest in farming as part of their identity and heritage. For example, the 26-year-old male university graduate, teacher and farmer from Idi Amu and Igbokiti, Nigeria, who was quoted in Chapter 8 (this volume), said he could not drop farming: 'It is built into me.'

Particularly in Tanzania and Zimbabwe, a number of young women and men who are not currently involved in agriculture imagined themselves starting their own farms. For example, a 22-year-old male in Zimbabwe who already owns a bar, plans to farm on land given to him by his mother. He said he has seen his neighbours succeed in farming and he wants to try it.

If farming is important in the imagined futures of many young people, in general, it is a very different kind of engagement with agriculture that they imagine. Particularly in Uganda, Ethiopia, Nigeria and Côte d'Ivoire these young people imagine that their future farming will involve little if any manual work ('digging'). Instead they imagine themselves in supervisory or managerial roles, with the day-to-day work being left to hired labour.

Not everyone imagined a farming future, not even as a manager of hired labour. For these individuals farming was too physically demanding, tiring or stressful, or the income was too little.

> I do not like farming. [...] I don't have the strength to do so. [...] The work is quite tiresome. When I return from work each day, I am so tired, and I would have to go the next day. [...] If I have nothing else to do, then I will go back to farming.
>
> (21-year-old male, Igbariam Eziafor, Nigeria)

> No, for the little I am doing, if I get a better thing going, I will leave it [farming]. There is no land here to use. The people of Igbariam community, like the people of Eket, do not have enough land to farm. Farm work! *Ana aputa aputa, ina abanye abanye.* [If people are running away from farming, and you are entering in to farming.] If opportunity comes, I will go to school; I don't want to go back to farm again.
>
> (25-year-old female, Igbariam Eziafor, Nigeria)

> Right now, if I have money, I'd buy rice in [large] quantities to keep it and then I'd sell it later, so that it helps me to move ahead in my trading. [...] Me, I want to establish business as my base... I want to stop working in the fields.
>
> (28-year-old female, Daloa, Côte d'Ivoire)

Young people who operated small businesses (many of whom were also involved in farming) imagined that these ventures would expand. Those who worked from street stalls or other informal structures, for example, imagined formalizing their businesses by moving into permanent and legal premises. This would make them less vulnerable to harassment by local officials, and would signal an important socio-economic achievement. In Tanzania, young people imagined expanding their businesses by either increasing the number of workers, the stock of merchandise, or opening up additional shops in surrounding areas. The fact that the Tanzania site, Mvurwi, is a thriving commercial centre and could provide a viable base for expansion into surrounding areas, likely informs these imaginaries. A managerial imaginary similar to that noted above for farming, was also evident in relation to a whole range of non-farm activities including shops, transport, building rental, catering and charcoal production.

> Um, in any case, I know that in the future, I won't do carpentry. [...] Well, when I say that, it's not that I won't have a workshop, but to say that I'll work as a carpenter, I don't think so. I have other projects in mind. [...] Well, doing commerce... when I say doing commerce, [I mean] a hardware store, and a small shop as well, if God wants it so.
> (28-year-old male, Daloa, Côte d'Ivoire)

For some individuals, future involvement in agriculture is articulated in terms of food security, identify and heritage, for others, imagined engagement in farming at a larger scale reflects ambitions to accumulate the financial resources to start or expand non-farm enterprises. The businesses reported in these imagined futures, including house painting, selling of building materials, trading in livestock, a mobile phone repair shop, a small restaurant, and constructing accommodation to rent out, are to be based in rural areas or in small towns.

The link between farm and non-farm economic activity is at the heart of the imagined futures of many young people. As a young man in Idi Amu, Nigeria, put it: 'Once the plantation is established, it's a cool source of income which can generate money to invest in other businesses' (24-year-old male). A young woman from Zimbabwe said:

> I expect to operate my own business. I expect to open my own takeaway [food] shop if I get money. All I need is a place, some plates and a refrigerator. I have no timeline in which I can achieve this, I just hope I can generate more money so that I will be able to operate my own business. The only challenge is to generate enough capital. To counter this challenge, I have ventured into farming this year. I planted my 1 ha of maize on my father's plot.
> (25-year-old female, Mvurwi, Zimbabwe)

> In the name of God, a good shopkeeper is better than a great farmer! A little shop on the side, with a bit of livestock keeping. [With] that [I will] sort myself out, because I've already lived a bit [out of] a shop; well I know how to go about it. Now, if I get a little problem at the shop, I'll use the livestock keeping to sort myself out.
> (29-year-old male, Korhogo, Côte d'Ivoire)

Migration is one aspect of diversification that figures prominently in the imagined futures of young people, particularly in Nigeria and Côte d'Ivoire (Table 9.2 and Appendix Tables 9A–D).

For some young people, migration is very much about accessing job opportunities and better services, and in this sense their imagined futures echo the dominant narrative that young people want to leave rural areas.

> Like in 2 years' time, as soon as I get enough money to pay rent for a shop and accommodation, I will leave this community.

Table 9.2. Importance of migration in young people's imagined futures.

Country	Site	Migration important in imagined future (%)
Uganda	Awach [relatively hot]	23
	Luweero [relatively hot]	41
	Mucwini [relatively cold]	5
	Butuntumula [relatively cold]	17
Ethiopia	Wondo Genet [relatively hot]	22
	Bora [relatively hot]	33
	Jabi Tehnan [relatively cold]	48
	Kuyu [relatively cold]	38
Nigeria	Oba Oke [relatively hot]	56
	Igbariam Eziafor [relatively hot]	38
	Idi Amu [relatively cold]	54
	Umumbo [relatively cold]	37
Côte d'Ivoire	Soubré [relatively hot]	71
	Odienné [relatively hot]	68
	Daloa [relatively cold]	54

> […] I no longer want to stay here. For instance, sewing clothes in this community is seasonal but when I migrate to an urban area, I will have clothes to sew throughout the year.
> (30-year-old female, Umumbo, Nigeria)

> Here is a village; we don't have much income-earning activities, there is no steady electricity; when you finish education, no job opportunities. The security situation in the country is bad. At least [in the USA] there will be steady supply of electricity and job opportunities.
> (25-year-old female, Igbariam Eziafor, Nigeria)

> In fact, I don't like remaining in the same place so much; it gives me a headache. I prefer to go somewhere – maybe return to Abidjan. Since I was in Abidjan before, I'll return to Abidjan.
> (25-year-old male, Daloa, Côte d'Ivoire)

> The city [Bouake, the second largest city in Côte d'Ivoire] being a bit big, one can have many social relations and do many things. That's what we imagine here, we can do many things over there. […] Here, it's going to the field that interests us. […] Umm, there aren't enough things in Zievasso, because here we can't, we can't even have social relations. We can't do anything if it's not working in the fields.
> (22-year-old male, Odienné, Côte d'Ivoire)

> […] It's Paris […] because it's pretty; […] it's the good life over there. […] There are apartment buildings, not mud huts, everything is built vertically; […] there are no fields there. […] Well, you live as you want […] you're comfortable. […] Because when I'll be there, I'll be respected […] when I'll come back here […] they'll say, she came back, the daughter of that person has arrived. […] No, no, she got a job there, she got hired […].
> (21-year-old female, Soubré, Côte d'Ivoire)

Some parents are supportive of these ambitions: 'There is no parent that will want their children to come back to this village except those who did not know where else to go again' (40-year-old female, Igbariam Eziafor, Nigeria).

For other young people, moving to and investing in nearby towns is part of an imagined future that continues to have a strong rural orientation, with agriculture continuing to play an important role. And still others, particularly in Nigeria, see migration as a way into more modern agriculture: 'People of this community like farming and also in Osun State as whole, but they do not practice mechanized farming like they do in Ilorin' (25-year-old male, Oba Oke, Nigeria).

Accumulating wealth or assets

Across all the sites in the seven study countries, both young women and men imagined futures in which they had accumulated valued assets such as land, farm inputs, cars, a building plot and housing. A number suggested that these assets represented 'development' or a 'civilised' life, while a 27-year-old man from Zimbabwe suggested that to be a real man you needed to have your own place, and not always be renting.[ii] The contrast between an asset-poor present and an imagined asset-rich future came through a photo–voice interview with a young man at the Umumbo site in Nigeria. Looking at a picture he took, he said:

> This is a picture of a beautiful house and a nice car. In that house, everything is complete including toilet and bathroom. The kind of person I feel in this picture is very wealthy and successful man that is very influential in the society. […] The reason why it's better is that currently I don't have money, I don't live in a good house; I live in a mud house built with a bamboo tree but this one [my imagined future] shows that development has come and that I have built a nice house.
> (19-year-old male, Umumbo, Nigeria)

The importance of building a house was mentioned repeatedly, due in part to the fact that many of the young people are currently living with a parent or a friend, or renting accommodation. Some young people imagine that owning a house would be less expensive than renting, and the savings could be invested in further economic activity. Many young married women live in their husbands' home, but imagined building their own dwelling to give them more security. One 30-year-old woman in Butuntumula, Uganda, mentioned that she was currently building a house for herself, of which her husband had no knowledge.

In Zimbabwe, young people imagined becoming landowners either through the Fast Track Land Reform Programme (FTLRP; see Scoones *et al.*, 2010) or by purchasing land. They explained that holding some form of land title could open doors – to credit or government subsidies for inputs, machinery, seeds and so on. Input schemes run by the Zimbabwean government, like Command Agriculture and the Presidential Input Scheme, only provide subsidies to farmers who have a 'paper' that formalizes their relationship to their land.

Further education and obtaining a professional or salaried job

As highlighted in Chapter 7 (this volume) many young people reported having to stop their formal education early because of financial constraints. Across all sites, additional education was an important part of many young people's imagined futures, although markedly less so in Ethiopia (Table 9.3 and Appendix Tables 9A–D). Overall, education was more likely to be important in the imagined futures of women, individuals who were younger, and those who already had relatively more education.

Young people – both those who have done well in school and many who had previously dropped out and imagine returning – regard education as instrumental for gaining professional or salaried employment. Jobs with government or private firms are particularly important in the imagined futures of those who are more educated or are working toward additional educational qualifications. The specific jobs mentioned included pharmacist, banker, secondary school teacher, veterinary nurse, gynaecologist, accountant, engineer and politician, among others.

Table 9.3. Importance of education in young people's imagined futures.

Country	Site	Education important in imagined future (%)
Uganda	Awach [relatively hot]	72
	Luweero [relatively hot]	33
	Mucwini [relatively cold]	50
	Butuntumula [relatively cold]	48
Ethiopia	Wondo Genet [relatively hot]	91
	Bora [relatively hot]	25
	Jabi Tehnan [relatively cold]	29
	Kuyu [relatively cold]	21
Nigeria	Oba Oke [relatively hot]	52
	Igbariam Eziafor [relatively hot]	69
	Idi Amu [relatively cold]	50
	Umumbo [relatively cold]	63
Côte d'Ivoire	Soubré [relatively hot]	67
	Odienné [relatively hot]	56
	Daloa [relatively cold]	62

> After I complete my education, I will be a government employee. [...] I [want to] serve as judge and in other positions. [...] I want to be a leader, a leader of the country. I will reach to a good status. I also think that I would be Prime Minister.
>
> (27-year-old male, Wondo Genet, Ethiopia)

> My uncle [is a lawyer]. [...] They dispense cases and justice for people. They are referred to as very learned people because they read a lot of books to be able to dispense judgement. They are highly regarded, with so much respect in the society.
>
> (19-year-old female, Oba Oke, Nigeria)

Education is the key that can open up such possibilities:

> Farming cannot offer me what education can offer. Educated people can be rich but farming is not like that. When you have graduated and obtained a certificate, then you start looking for a white-collar job.
>
> (20-year-old female, Kuyu, Ethiopia)

> Schooling is more important because it will give me a continuing source of livelihood, it does not fade. You can always use your certificate to get any good job at any time. Farming, it can fade. If you do it and you are ageing, you might not be fit to continue with farming. But with your education and certificate you can get paid work and have some money, you can hire people to do farm work for you. That is the reason why schooling is more important.
>
> (25-year-old female, Oba Oke, Nigeria)

The advantages of white-collar employment were contrasted with the demands of farming:

> Education can make you pensionable when you grow old while farming will not. It can also make you train your children in school and other businesses in life. In the case of death, maybe when one's husband dies, one who is a

> civil servant can use money from her salary to train her children, but farming is not like. Farming gets old once the owner is old and these are the reasons education is good.
>
> (18-year-old female, Oba Oke, Nigeria)

In contrast, for individuals who are a few years older, or who have already started a family, the focus tended to be on vocational training that would allow them to move toward an imagined future in which they raise their current activities to a new level, or move into new activities.

> Well maybe some vocational training in salon work would improve on what I already know. However, the major constraint is lack of money to do this. My responsibilities are so many, and right now my children's education is a priority over mine.
>
> (34-year-old female, Awach, Uganda)

> The work that I feel I should be doing since I am growing older [...] if I can get some strength or ability I should get a training in carpentry or building construction.
>
> (32-year-old male, Awach, Uganda)

Among those who imagined securing professional work through education, some still saw their future selves being involved in farming and the non-farm rural economy. For example, one young woman was studying computing but had a family background in farming. She imagined working as a civil servant, having a hairdressing salon, and also farming:

> Even if I have all the money in the world I cannot stop farming. I will still consider farming, but this time more mechanized farming and I will use fertilizer to increase yield.
>
> (28-year-old female, Oba Oke, Nigeria)

Another who was training to be a teacher said that alongside teaching she imagined herself specializing in rice cultivation, because of its returns:

> I would like to combine my work with rice cultivation, to make ends meet and help my family. [...] I would like to stop farming other crops because there is lots of stress in farming other crops, while they don't give many returns.
>
> (21-year-old female, Mucwini, Uganda)

In Nigeria many young people imagined combining a salaried job with larger scale mechanized farming that was carried out away from their respective communities which had little available land. Others imagined giving the land out on a sharecropping basis. A young man from Umumbo, Nigeria, who was presently studying biochemistry at degree level, imagined himself employed as a biochemist and developing a business career, while also investing in farming.

Closely linked to the imaginary of furthering their own education is a vision of the future in which these young people can ensure the education of their own children. For those who already have children, educating them is now a higher priority than furthering their own education. The considerable importance of educating offspring in the imagined futures of the young people is to ensure that their children do not have to toil and 'suffer' in the same ways that they have. One woman from Nigeria explained why she was not encouraging her children to follow her into farming:

> Hahahaaa! No, I'm not. I'm only trying to put them in civil service. [...] I sponsor them to be educated and that will qualify them to get white-collar job as far as God is in control.
>
> (42-year-old female, Umumbo, Nigeria)

A young woman from Ethiopia said:

> We rather focus on educating our kids, now it is too late for us to go to school at this age. If we both go to school, who will do our jobs on which our family livelihood depends? Education and work do not go together; you have to sacrifice one to achieve the other. It is impossible to do both of them together.
>
> (28-year-old female, Bora, Ethiopia)

Moving from the present to the future

To move towards the futures they imagine for themselves, young women and men must navigate a variety of obstacles, of which the most commonly cited was a lack of financial capital or credit. As noted in Chapter 7 (this volume), across all the countries and sites, the absence of immediately available funds was the most common explanation given by young people for terminating their formal education. It was also a block to the additional education and training that is part of many imagined futures.

The issue here goes beyond a simple lack of cash. As explained by a young man in Ethiopia, re-entering education can affect ongoing income-generating activities, and thus result in a double financial burden:

> Education is very important. I do not have anything hindering my decision except the issue of money, because education is too expensive and I fear losing my customers in my orange business.
>
> (23-year-old male, Kuyu, Ethiopia)

This extra burden is expected to be particularly important for those who are already responsible for children, siblings or ageing parents. Some young people acknowledged that, due to financial constraints, they would likely have to adapt their plans for further education in order to manage the financial burden.

> Even if you get financial support it is not convenient to work and learn at the same time. Because you have to learn during the whole day, you cannot get money to pay for house rent and other expenses even if you get some support. So, to pursue education there should be at least be full support, but that is not possible. It is better to work with the education you have obtained.
>
> (23-year-old male, Kuyu, Ethiopia)

The challenge of finance goes far beyond education, and in one way or another it permeates almost all young people's discussions of their futures. A 27-year-old man in Zimbabwe, for example, complained that despite having a dam close to their farm, they could not afford to invest in irrigation equipment, purchase inputs, and 'upgrade' into fully mechanized commercial farming. A 35-year-old male migrant who had settled in Tuobodom, Ghana, and who engaged in commercial tomato farming, imagined buying cows and goats in the north and transporting these to Accra for onward sale. However, he acknowledged that without finance this would not be possible. A 25-year-old man from the same location who was working as a farm labourer, imagined becoming a professional driver in the near future, although he was quick to recognize the need to first gather resources to secure a driving licence. A 26-year-old man in Awach, Uganda stated:

> Electricity supply can limit me in the welding business because it might require a lot of money to sort it out. A building to put the machines can also be a problem, and also acquiring the grinding and cutting machines.

We recognize that from the young people's perspective, pleading poverty (or lack of financial liquidity) in an interview with researchers might be easy, or perhaps even expected. It is also clear that in many situations, between ongoing economic activities, inputs from family and participation in savings groups, small amounts of capital are often available to start, maintain or expand farming or other activities (Flynn and Sumberg, 2017, 2018). Nevertheless, the young people are pointing to an important and pervasive tension between discourses highlighting youth entrepreneurship, innovation and 'farming as a business' (see Chapter 6, this volume) on the one hand, and young people's experiences of constrained access to financial resources on the other.

In addition to finance, for some young people, moving toward their imagined futures means navigating gender and other social norms, which are linked to broader narratives about women and men. For example, in Uganda there was a strong sense among young men that women focus on marriage and men on financial independence; while in the eyes of young women generally, men are seen to be smoking, drinking and gambling, while they themselves are more focused on the future.

However, as is often the case, there is a gap between cited norms and actual lives and livelihoods. For example, some young people in Uganda observed that 'there are women who think like men and have the same vision as men, and they focus on all kinds of work including those normally reserved for men'. Even though in all sites some jobs are seen to be 'for men', one woman in Uganda observed: 'If I study or learn how to do it, then I could try and challenge the men at this job.' A young woman from Jabi Tehnan village, Ethiopia, was less sanguine:

> The place given to women is not that good. I mean that people in this area do not adequately support the expectations that women have. They feel women do not properly accomplish things. I am encouraged to do activities because of the support from my parents and my brothers. Other people start rumours about me, that I am going to do this or that activity. Seeing women as inferior and unable to accomplish a certain goal is prevalent in this area. I think this perception may hinder the implementation of my plans.
>
> (26-year-old female, Jabi Tehnan, Ethiopia)

A final set of hurdles identified by young people might best be described as hazard or idiosyncratic shocks, including drought, illness, death and unexpected family obligations, as well as theft and customers not honouring their debts. A 20-year-old man in Tanzania explained that his money goes to helping his parents, and this hinders him from moving towards his imagined future. This is the other side of the coin of the support from family and friends, from which so many of the other young people reported benefiting. In Zimbabwe, two producers (26-year-old female and 27-year-old male) spoke about weather and climate change as important challenges to their future agricultural prospects. Without the ability to irrigate, they expect farming to become increasingly risky. One of these individuals recently experienced drought in one year and heavy rain in another (the El Niño and La Niña phenomena) which resulted in a major setback.

It is clear, and it was often remarked upon by the young people, that for many of them movement toward their imagined futures is likely to be slow and difficult. Some lack the capital and strong social networks, and in some ways their situation resembles certain aspects of waithood. Others are already on their way: they remain positive, stressing the importance of hard work, focus and determination, savings, and strong social networks.

Discussion and Conclusions

Perhaps it should not be a surprise that young people who are centrally engaged in the rural economy, often by combining agriculture and non-farm work (see Chapter 3, this volume; Yeboah *et al.*, 2020), imagine futures for themselves that are still largely rural, and in which agriculture continues to play an important part. Nevertheless, at least in some locations, these imagined futures pose a challenge to the orthodoxy that large numbers of African youth cannot wait to leave the countryside. This does not however mean that all is well among young people in rural SSA, or that the promise of the 'rural prosperity gospel' is finally coming good.

It is true that many young women and men imagine futures for themselves in which agriculture continues to play an important part (see also Kristensen and Birch-Thomsen, 2013; Berckmoes and White, 2014; Daum, 2019; Yeboah *et al.*, 2020). Yet this should not be interpreted simply as young people actively choosing 'to be farmers' or to have 'a career in farming'. Their imagined futures, while largely rural and small-town focused, contain two important and related wrinkles in relation to agriculture. First, in the vast majority of these futures, farming and/or livestock is combined with other, non-farm economic activities, from salaried jobs through to catering, petty trading and brick making. The young people are envisaging what have been called 'mixed', 'portfolio' or 'diversified' livelihoods: but strikingly, it is not clear whether agriculture is the mainstay or the 'side-hustle' (Mwaura, 2017). In other words, while farming is part of the picture, it is most often not as the 'full-time farmer' whose image and presumed motivations inform much agricultural policy and underpin the assumptions of both agricultural research and development programmes (Sumberg *et al.*, 2004).

Second, in many imagined futures the engagement with agriculture is as manager or supervisor of hired labour. As imagined by some young people, this would amount to farming at arm's length by a busy, town-based manager of multiple enterprises. In a sense these young people are aligning their futures with current discourses around 'telephone farmers' (Leenstra, 2014) and 'medium-scale farms' (Jayne *et al.*, 2016, 2019). In Nigeria, the common reference to 'large-scale mechanized farming' by young people points in the same direction. Again, this imaginary of future farming raises fundamental questions in relation to many aspects of agricultural and rural development, such as the role of labour-saving technologies and the likely implications for investment in alternative farming systems (e.g. low maintenance tree crops versus food crops) and other forms of productivity-enhancing technology.

Formal education – its poor quality, being forced to leave early, and the strong desire to do more – is the backdrop against which both current activities are undertaken and futures are imagined. Despite the many disappointments relating to formal education, the idea of continuing or restarting education or vocational training figures prominently in many imagined futures. For many young people this is about joining and succeeding in the modern, formal world of work, and enjoying the associated financial and social rewards (see also Bajema *et al.*, 2002; Leavy and Smith, 2010). The ways in which government and wage work are valued by the young people who participated in the research reflects a hierarchy of respect and prestige accorded to different kinds of professional work, most often found in urban areas, and a dissatisfaction with their current status in the rural labour market. Again, this dissatisfaction does not generally translate into imagined futures that are non-rural or non-agricultural. Indeed, some young people clearly articulated links between further education, wage employment, and improved agricultural performance, better care of livestock and so on.

Current economic activities allow some young people to accumulate assets including land, housing and vehicles, and such assets also figure prominently in the futures that young people more generally imagine for themselves. How well the diversified, farming-inclusive rural livelihoods they envisage will support this level of asset accumulation, and who among the young people will have the financial and other resources to make these portfolios work to their advantage, are open questions. These queries are open, but extremely important, as they have direct implications for attainment of social adulthood and thus for the discussion of waithood (see Chapter 8, this volume), as well as for the dynamics of African agricultural and rural development into the future.

Finally, in this analysis of the futures that young people imagine for themselves, we were not able identify any consistent differences associated with categorization of the economic dynamism of the different sites, as either relatively hot or relatively cold. This could be because of the relatively small sample, or because a hot/cold binary classification is not sufficient to capture the multiple dimensions and complexities of local economic dynamism. Alternatively, it may be that what appear on the surface as importance differences in economic dynamism do not actually correspond to different perceptions on the part of young people, of either present or future landscapes of opportunity. In any case, this is an area that certainly deserves more research attention.

Notes

[i] The Agricultural Policy Research on Africa (APRA) is a 5-year programme, funded by the UK Department for International Development (DFID), focused on pathways to agricultural commercialization. See: https://www.future-agricultures.org/apra/ (accessed 26 January 2021).

[ii] '*Kuti uve murume unofanirwa kuvanemusha wako, haikono kugara uchilodger.*' [To be a real man you needed to have your own place, and not always be renting] (27-year-old male, Zimbabwe).

References

Andersson Djurfeldt, A., Kalindi, A., Lindsjö, K. & Wamulume, M. 2019. Yearning to farm – Youth, agricultural intensification and land in Mkushi, Zambia. *Journal of Rural Studies,* 71, 85–93.

Anyidoho, N. A., Leavy, J. & Asenso-Okyere, K. 2012. Perceptions and aspirations: A case study of young people in Ghana's cocoa sector. *IDS Bulletin,* 43, 20–32.

Bajema, D. H., Wade Miller, W. & Williams, D. 2002. Aspirations of rural youth. *Journal of Agricultural Education,* 43, 61–71.

Ball, S. J., Macrae, S. & Maguire, M. E. G. 1999. Young lives, diverse choices and imagined futures in an education and training market. *International Journal of Inclusive Education,* 3, 195–224.

Berckmoes, L. & White, B. 2014. Youth, farming and precarity in rural Burundi. *European Journal of Development Research,* 26, 190–203.

Bernard, T. & Seyoum Taffesse, A. 2014. Aspirations: An approach to measurement with validation using Ethiopian data. *Journal of African Economies,* 23, 189–224.

Bessant, J., Pickard, S. & Watts, R. 2020. Translating Bourdieu into youth studies. *Journal of Youth Studies,* 23, 79–92.

Bourdieu, P. 1984. *Distinction: A Social Critique of the Judgement of Taste*, London, Routledge.

Bourdieu, P. & Wacquant, L. J. D. 1992. *An Invitation to Reflexive Sociology,* Cambridge, UK, Polity Press.

Daum, T. 2019. Of bulls and bulbs: Aspirations, opinions and perceptions of rural youth in Zambia. *Development in Practice,* 29, 882–897.

Favara, M. 2017. Do dreams come true? Aspirations and educational attainments of Ethiopian boys and girls. *Journal of African Economies,* 26, 561–583.

Flynn, J. & Sumberg, J. 2017. Youth savings groups in Africa: they're a family affair. *Enterprise Development and Microfinance,* 28, 147–161.

Flynn, J. & Sumberg, J. 2018. Are savings groups a livelihoods game changer for young people in Africa? *Development in Practice,* 28, 51–64.

Gastineau, B. & Golaz, V. 2016. Being young in rural Africa. Introduction. *Afrique Contemporaine,* 259, 9–22.

Hardgrove, A., Rootham, E. & Mcdowell, L. 2015. Possible selves in a precarious labour market: youth, imagined futures, and transitions to work in the UK. *Geoforum,* 60, 163–171.

Jayne, T. S., Chamberlin, J., Traub, L., Sitko, N., Muyanga, M., Yeboah, F. K., Anseeuw, W., Chapoto, A., Wineman, A. & Nkonde, C. 2016. Africa's changing farm size distribution patterns: the rise of medium-scale farms. *Agricultural Economics,* 47, 197–214.

Jayne, T. S., Muyanga, M., Wineman, A., Ghebru, H., Stevens, C., Stickler, M., Chapoto, A., Anseeuw, W., Van Der Westhuizen, D. & Nyange, D. 2019. Are medium-scale farms driving agricultural transformation in sub-Saharan Africa? *Agricultural Economics,* 50, 75–95.

Kosec, K., Khan, H., Taffesse, A. S. & Tadesse, F. 2014. Promoting aspirations for resilience among the rural poor. *In:* Fan, S., Pandya-Lorch, R. & Yosef, S. (eds.) *Resilience for Food and Nutrition Security,* Washington, DC, International Food Policy Research Institute (IFPRI).

Kristensen, S. B. P. & Birch-Thomsen, T. 2013. Should I stay or should I go? Rural youth employment in Uganda and Zambia. *International Development Planning Review,* 35, 175–201.

Leavy, J. & Hossain, N. 2014. *Who Wants to Farm? Youth Aspirations, Opportunities and Rising Food Prices. IDS Working Paper 439,* Brighton, UK, Institute of Development Studies (IDS).

Leavy, J. & Smith, S. 2010. *Future Farmers: Youth Aspirations, Expectations and Life Choices. FAC Discussion Paper 013,* Brighton, Future Agricultures Consortium.

Leenstra, M. 2014. From suitcase farmers to telephone farmers. Agriculture and diversified livelihoods among urban professionals. *In:* Akinyoade, A., Klaver, W., Soeters, S. & Foeken, D. (eds.) *Digging Deeper: Inside Africa's Agricultural, Food and Nutrition Dynamics,* Leiden, the Netherlands, Brill.

Locke, C. & te Lintelo, D. 2012. Young Zambians 'waiting' for opportunities and 'working towards' living well: lifecourse and aspiration in youth transitions. *Journal of International Development,* 24, 777–794.

Markus, H. & Nurius, P. 1986. Possible selves. *American Psychologist,* 41, 954–969.

Mausch, K., Harris, D., Heather, E.E.J., Yim, J. & Hauser, M. 2018. Households' aspirations for rural development through agriculture. *Outlook on Agriculture,* 47, 108–115.

McKendrick, L. A. C. J. H. 2017. Beyond aspirations: deploying the capability approach to tackle the under-representation in higher education of young people from deprived communities. *Studies in Continuing Education,* 39, 120–137.

Moore, K. 2015. Fostering economic opportunities for youth in Africa: a comprehensive approach. *Enterprise Development and Microfinance,* 26, 195–209.

Mwaura, G. M. 2017. The side-hustle: diversified livelihoods of Kenyan educated young farmers. *IDS Bulletin,* 48, 51–66.

OECD 2017. *Youth Aspirations and the Reality of Jobs in Developing Countries: Mind the Gap,* Paris, Development Centre Studies, OECD Publishing.

Petesch, P. & Rodríguez Caillava, I. 2012. *Voices of Young Villagers in Sub-Saharan Africa. Background paper,* Washington, DC, The World Bank.

Scoones, I., Marongwe, N., Mavedzenge, B., Mahenehene, J., Murimbarimba, F. & Sukume, C. 2010. *Zimbabwe's Land Reform: Myths & Realities*, Oxford, UK, James Currey.

Skeggs, B. 2011. Imagining personhood differently: person value and autonomist working-class value practices. *The Sociological Review,* 59, 496–513.

Stahl, G., Mcdonald, S. & Young, J. 2020. Possible selves in a transforming economy: Upwardly mobile working-class masculinities, service work and negotiated aspirations in Australia. *Work, Employment and Society,* (doi.org/10.1177/0950017020922336).

Sumberg, J., Gilbert, E. & Blackie, M. 2004. Income diversity, technology choice and agricultural research policy in sub-Saharan Africa. *Development Policy Review,* 22, 131–146.

Sumberg, J., Yeboah, T., Flynn, J. & Anyidoho, N. A. 2017. Young people's perspectives on farming in Ghana: a Q study. *Food Security,* 9, 151–161.

Verkaart, S., Mausch, K. & Harris, D. 2018. Who are those people we call farmers? Rural Kenyan aspirations and realities. *Development in Practice,* 28, 468–479.

White, B. 2012. Agriculture and the generation problem: rural youth, employment and the future of farming. *IDS Bulletin,* 43, 9–19.

White, B. 2019. *Rural Youth, Today and Tomorrow. IFAD Research Series 48*, Rome, International Fund for Agricultural Development (IFAD).

Yeboah, T., Sumberg, J., Flynn, J. & Anyidoho, N. A. 2017. Perspectives on desirable work: Findings from a Q study with students and parents in rural Ghana. *European Journal of Development Research,* 29, 423–440.

Yeboah, T., Chigumira, E., John, I., Anyidoho, N. A., Manyong, V., Flynn, J. & Sumberg, J. 2020. Hard work and hazard: Young people and agricultural commercialisation in Africa. *Journal of Rural Studies,* 76, 142–151.

Appendix

Table 9A. Uganda comparison across sites: imagined futures (IF).[a]

	Sex		Age				Education[b]				
	F	M	18–21	22–25	26–29	30–33	P	P+	S	S+	Overall
Awach [relatively hot]											
Agriculture is important in IF (%)	93	92	91	100	80	100	100	89	89	88	92
Non-farm is important in IF (%)	100	88	100	100	90	75	92	100	89	88	92
Migration is important in IF (%)	14	28	18	44	0	25	15	22	11	50	23
Education is important in IF (%)	86	64	91	78	70	38	62	78	78	75	72
N (individuals)	14	25	11	9	10	8	13	9	9	8	39
Luweero [relatively hot]											
Agriculture is important in IF (%)	72	86	86	60	87	89	75	80	84	70	78
Non-farm is important in IF (%)	96	86	100	87	93	89	92	100	84	100	91
Migration is important in IF (%)	56	24	57	33	27	67	42	20	37	60	41
Education is important in IF (%)	36	29	57	53	13	11	17	40	42	30	33
N (individuals)	25	21	7	15	15	9	12	5	19	10	46
Mucwini [relatively cold]											
Agriculture is important in IF (%)	94	83	69	100	93	100	87	100	92	71	88
Non-farm is important in IF (%)	75	88	100	78	73	67	67	83	100	86	83
Migration is important in IF (%)	0	8	8	11	0	0	0	0	8	14	5
Education is important in IF (%)	38	58	77	67	20	33	33	17	83	57	50
N (individuals)	16	24	13	9	15	3	15	6	12	7	40
Butuntumula [relatively cold]											
Agriculture is important in IF (%)	67	64	71	50	60	80	67	100	50	75	65
Non-farm is important in IF (%)	100	100	100	100	100	100	100	100	100	100	100
Migration is important in IF (%)	8	27	29	33	0	0	33	0	30	0	17
Education is important in IF (%)	67	27	57	83	20	20	0	0	70	50	48
N (individuals)	12	11	7	6	5	5	3	2	10	8	23

[a]Column totals exceed 100% because multiple options were flagged as important in the respondent's imagined futures.
[b]P = some primary education; P+ = some primary education plus some additional training; S = some secondary education; S+ = some secondary education plus some additional training.

Table 9B. Ethiopia comparison across sites: imagined futures (IF).[a]

	Sex		Age				Education[b]					
	F	M	18–21	22–25	26–29	30–33	None	P	P+	S	S+	Overall
Wondo Genet [relatively hot]												
Agriculture is important in IF (%)	17	55	13	17	71	–	–	50	0	30	40	35
Non-farm is important in IF (%)	100	100	100	100	100	–	–	100	100	100	100	100
Migration is important in IF (%)	33	9	50	17	0	–	–	50	0	30	10	22
Education is important in IF (%)	92	91	88	100	86	–	–	100	100	80	100	91
N (individuals)	12	11	8	6	7	–	–	2	1	10	10	23
Bora [relatively hot]												
Agriculture is important in IF (%)	62	82	44	100	88	83	50	85	50	50	100	71
Non-farm is important in IF (%)	92	55	89	0	75	67	50	69	50	100	100	75
Migration is important in IF (%)	15	55	22	100	25	50	50	38	50	17	0	33
Education is important in IF (%)	31	18	44	0	13	17	0	23	50	17	100	25
N (individuals)	13	11	9	1	8	6	2	13	2	6	1	24
Jabi Tehnan [relatively cold]												
Agriculture is important in IF (%)	30	73	0	100	43	67	100	50	67	38	100	52
Non-farm is important in IF (%)	100	73	100	80	86	67	100	50	67	92	100	86
Migration is important in IF (%)	40	55	0	40	57	100	100	50	67	38	50	48
Education is important in IF (%)	0	55	25	40	14	67	0	0	67	23	50	29
N (individuals)	10	11	4	5	7	3	1	2	3	13	2	21
Kuyu [relatively cold]												
Agriculture is important in IF (%)	42	33	38	25	100	33	0	57	50	30	25	38
Non-farm is important in IF (%)	92	100	100	88	100	100	100	86	100	100	100	96
Migration is important in IF (%)	33	42	25	50	100	17	100	14	0	50	50	38
Education is important in IF (%)	25	17	38	13	0	17	0	14	0	20	50	21
N (individuals)	12	12	8	8	2	6	1	7	2	10	4	24

[a]Column totals exceed 100% because multiple options were flagged as important in the respondent's imagined futures.
[b]P = some primary education; P+ = some primary education plus some additional training; S = some secondary education; S+ = some secondary education plus some additional training.

Table 9C. Nigeria comparison across sites: imagined futures (IF).[a]

	Sex		Age				Education[b]					
	F	M	18–21	22–25	26–29	30–33	Train	P	P+	S	S+	Overall
Oba Oke [relatively hot]												
Agriculture is important in IF (%)	23	57	27	55	40	–	–	100	50	67	28	41
Non-farm is important in IF (%)	100	86	100	91	80	–	–	100	100	100	89	93
Migration is important in IF (%)	62	50	64	55	40	–	–	100	0	67	56	56
Education is important in IF (%)	62	43	82	36	20	–	–	100	0	33	61	52
N (individuals)	13	14	11	11	5	–	–	1	2	6	18	27
Igbariam Eziafor [relatively hot]												
Agriculture is important in IF (%)	38	38	50	40	33	0	0	0	50	60	33	38
Non-farm is important in IF (%)	100	100	100	100	100	100	100	0	100	100	100	100
Migration is important in IF (%)	46	31	50	40	17	100	100	0	0	20	44	38
Education is important in IF (%)	77	62	100	80	17	100	100	0	50	60	72	69
N (individuals)	13	13	4	15	6	1	1	0	2	5	18	26
Idi Amu [relatively cold]												
Agriculture is important in IF (%)	38	100	78	73	40	100	100	100	0	67	74	69
Non-farm is important in IF (%)	100	100	100	100	100	100	100	100	100	100	100	100
Migration is important in IF (%)	69	38	67	55	40	0	0	100	100	67	47	54
Education is important in IF (%)	38	62	78	27	40	100	100	0	0	67	53	50
N (individuals)	13	13	9	11	5	1	1	1	2	3	19	26
Umumbo [relatively cold]												
Agriculture is important in IF (%)	62	57	44	60	75	75	–	100	100	33	64	59
Non-farm is important in IF (%)	85	86	89	80	75	100	–	50	100	89	86	85
Migration is important in IF (%)	31	43	44	10	75	50	–	0	0	33	50	37
Education is important in IF (%)	54	71	78	70	75	0	–	0	0	78	71	63
N (individuals)	13	14	9	10	4	4	–	2	2	9	14	27

[a]Column totals exceed 100% because multiple options were flagged as important in the respondent's imagined futures.
[b]Train = Training only; P = some primary education; P+ = some primary education plus some additional training; S = some secondary education; S+ = some secondary education plus some additional training.

Table 9D. Côte d'Ivoire comparison across sites: imagined futures (IF).[a]

	Sex		Age				Education[b]					
	F	M	18–21	22–25	26–29	30–33	None	P	P+	S	S+	Overall
Soubré [relatively hot]												
Agriculture is important in IF (%)	27	46	22	43	40	67	50	40	50	27	50	38
Non-farm is important in IF (%)	100	100	100	100	100	100	100	100	100	100	100	100
Migration is important in IF (%)	73	69	78	71	40	100	50	60	0	91	100	71
Education is important in IF (%)	73	62	78	71	60	33	100	40	50	73	50	67
N (individuals)	11	13	9	7	5	3	4	5	2	11	2	24
Odienné [relatively hot]												
Agriculture is important in IF (%)	38	46	31	50	50	100	20	60	100	50	50	44
Non-farm is important in IF (%)	100	92	100	100	100	50	100	100	100	83	100	96
Migration is important in IF (%)	62	77	77	63	50	50	70	60	50	83	50	68
Education is important in IF (%)	62	46	69	38	50	0	70	40	0	67	50	56
N (individuals)	13	13	13	8	2	2	10	5	2	6	2	25
Daloa [relatively cold]												
Agriculture is important in IF (%)	9	40	0	60	18	50	0	36	0	29	50	27
Non-farm is important in IF (%)	91	87	100	100	73	100	100	73	0	100	100	88
Migration is important in IF (%)	55	53	60	100	36	50	50	64	0	43	50	54
Education is important in IF (%)	45	73	60	60	64	75	67	36	0	86	100	62
N (individuals)	11	15	5	5	11	4	6	11	0	7	2	26

[a]Column totals exceed 100% because multiple options were flagged as important in the respondent's imagined futures.
[b]P = some primary education; P+ = some primary education plus some additional training; S = some secondary education; S+ = some secondary education plus some additional training.

10 Young People and the Rural Economy: Syntheses and Implications

James Sumberg[1], Carolina Szyp[1], Thomas Yeboah[2], Marjoke Oosterom[1], Barbara Crossouard[3], Jordan Chamberlin[4]
[1]Institute of Development Studies, Brighton, UK; [2]Bureau of Integrated Rural Development (BIRD), Kwame Nkrumah University of Science and Technology (KNUST), Kumasi, Ghana; [3]University of Sussex, Brighton, UK; [4]International Maize and Wheat Improvement Center (CIMMYT), Nairobi, Kenya

Introduction

The research presented in this book uses qualitative and quantitative analysis to address the dominant narratives and 'conventional wisdom' about youth and the rural economy in sub-Saharan Africa (SSA) as they have been configured in recent academic and policy literature. Throughout, we have sought to carefully consider the conceptualizations embedded in these narratives, their empirical foundations, and their implications for policy. Our findings lend support to some elements of the standard stories, while they challenge others. However, even where we find broad support for dominant narratives, we conclude that there is need for more nuance than is generally offered in shorthand treatments of 'youth' questions. This critical revisiting of the storytelling around youth in rural Africa has important ramifications for policy framing and discourse, as well as policy content. This final chapter synthesizes the empirical findings described in the previous chapters and sets out their discursive and practical implications for policy relating to youth, agricultural and rural development.

Synthesis

The broad story of contemporary rural youth in SSA that emerges from the research is one of livelihood building under severe and persistent constraints. Having been buffeted as children by forces beyond their control – including widespread poverty, parental illness or death, family break-up or civil conflict – young women and men are then let down by formal education. The quality of education provision is low, and many are forced to leave school early because it is simply not affordable. This is despite having worked, often from an early age, to help pay their school fees and support their households. Although many young people see it as normal to combine school and work, others recognize that this jeopardizes their educational progress. The deeply gendered rural opportunity landscapes they encounter offer few prospects for remunerative, secure or decent work, to say nothing of salaried employment. However, through their own hard work and with the support of their families and social networks, these young people set about to build their livelihoods, and in some cases accumulate assets, in contexts where infrastructure

DOI: 10.1079/9781789245011.0010

is poor and services lacking, and where gendered social norms and strong social hierarchies restrict room for manoeuvre, particularly for women.

These livelihoods reflect shifting patterns of engagement with the rural economy, combining unpaid caring and domestic work with farming, non-farm wage employment and/or non-farm self-employment. The informal and seasonal nature of much of this economic activity has given rise to endemic precarity, where work is characterized by risk, limited financial reward, instability and lack of protection (Sumberg *et al.*, 2020). Livelihood building extends well beyond the labour market, however, with young people navigating the challenges of securing accommodation and land, furthering their own education, caring for parents and siblings, and negotiating relationships, marriage, children and citizenship, as they strive for social adulthood. The futures they imagine for themselves usually involve expansion of their current activities and/or diversification into other employments, often including larger-scale, more modern agriculture. In many of these imagined futures young people are farming and running their businesses as managers of hired labour. Many also imagine restarting or furthering their education to boost their chances of securing a professional wage job and/or improving agricultural productivity. Mobility and migration also figure in many imagined futures: in some contexts, the focus is on nearby rural towns with the idea of maintaining a firm base in the rural economy, while in others it is imagined as the more classic flight to larger urban centres.

In other words, for the vast majority of young people in the study sites, their engagement with the rural economy in the early stages of livelihood building is best characterized as hard work in the face of hazard – personal, financial and environmental. However, the broad outlines of the story above are not universal, and we acknowledge that while the opportunity landscape is limited in most places, it is not equally limited everywhere. There are real spatial differences in opportunities (Abay *et al.*, 2020), as well as socially constructed and/or mediated differences. Drawing on Roberts (1968, 2009) our broad-brush picture supports the proposition that the job (and broader livelihood) opportunities available to rural young people in SSA emerge from multiple 'opportunity structures' that act to create distinct routes into the labour force and/or social adulthood. Indeed, these opportunity structures, emerging from a web of determinants including place, family origins, gender, ethnicity, education and labour market processes, were a central concern of the research.

A fundamental insight emerging from opportunity structure theory is that neither poor young people, nor poor adults, typically choose their jobs in any meaningful sense: 'they simply take what is available' (Roberts, 1977, p.3). This raises important questions about the preoccupation with youth aspirations in the development literature, and with the individual choices and decisions of young people in relation to work and livelihoods. The point is certainly not that everything is predetermined, but rather that most young people in rural SSA actually have relatively little room to manoeuvre. Thus, while some profess a deep attachment to farming, for many others it is the obvious (and perhaps only) 'choice' that allows them to assure food security, earn some income, and forge a potential path to social adulthood. Similarly, the 'choice' between selling charcoal or selling dried fish in the market, or between making bricks or doing day labour, is not irrelevant, but it is unlikely to result in significantly altered financial or social outcomes. While some imagine prosperous futures, where they are running (sometimes several) successful businesses and/or farms, or are engaging in white-collar professional work and are accumulating considerable material wealth and social status, it remains to be seen whether they can transform these imagined futures into reality.

Our intention is not to reify the notion of opportunity structures. Indeed, apart from what is obvious – better-off young people generally have more options; women and men have different options; migrants generally have fewer options (although this depends on their networks); and areas of higher potential offer more options than areas of lower potential – we see relatively little indication that opportunity structures work to finely differentiate how young people engage with the rural economy. Rather, within and between sites, and across an array of social variables (gender, age, education and so on) there are very strong similarities in the ways in which young people engage with and affect the rural economy. Most young people combine some farming with one or more other common, low-skill, low-investment, low-technology and low-return economic activities. This lack of diversity in patterns of engagement with the rural economy reflects a severely depleted opportunity landscape – resulting from poor infrastructure, limited purchasing power, poor policy and so on – as opposed to any generalized lack of ambition, skill or capital among young people. In such contexts, it is not surprising that young people are not the innovative drivers of change in the farm or non-farm economy, even if this is how they are often portrayed (cf. Sumberg and Hunt, 2019).

Implications

The broad synthesis outlined in the preceding section, reflecting field work in 21 sites across seven SSA countries, and as elaborated in the chapters of this book and other recent publications (Sumberg and Hunt, 2019; Sumberg *et al.*, 2019; Abay *et al.*, 2020; Carreras *et al.*, 2020; Chamberlin and Sumberg, 2020; Crossouard *et al.*, 2020; DeJaeghere, 2020; Glover and Sumberg, 2020; Oosterom *et al.*, 2020; Yeboah *et al.*, 2020) suggests that it is now time to reconsider both the framings and narratives that underpin policy, as well as policy content. In the sections that follow we explore the implications for policy framing and discourse, policy content, research and development practice.

Issues framing and discourse

Perhaps the most obvious, but also the most far reaching, implication of the outcomes of this research is the urgent need to reframe the 'problem' of Africa's rural youth, so that it is no longer 'all about youth' and their individual and collective deficits (lack of skills, lack of interest in hard work, lack of interest in education and so on). An alternative framing is now required that puts the economy and its inability to provide decent employment (for young people and all rural residents) centre stage (Sumberg *et al.*, 2020). This framing must emphasize the need for structural change, and link this directly to an appreciation of the central role of opportunity structures – from oppressive gender norms, failed education policy initiatives and patriarchal–gerontocratic local institutions, to poor infrastructure – in shaping young people's livelihood trajectories and outcomes. In making explicit the embeddedness of young people in broader networks of social, economic and political relations, such a framing will foreground interventions meant to promote structural change in the broadest sense, and will nudge the current generation of low-impact, youth-specific projects aside.

A second implication is that (actual or threatened) mass out-migration by rural young people must be dislodged as a core element of policy discourse. Yes, some young people want to – and do – leave; in some locations, this might have demographic significance. However, millions and millions of young people keep one or both feet in rural areas as they move on in life, in pursuit of better livelihood and educational opportunities, and social status. The fact that for many, neither their current economic activities, nor their imagined futures reflect the archetypical image of a full-time 'family farmer' is irrelevant, except in so much as this outdated image continues to underpin problem framing and policy.

Similarly, more care is needed in the way the notion of 'waithood' is being uncritically integrated into policy and public discourse. Specifically, the claim that the majority of young people are stuck in permanent waithood is far too broad, while the supposed link between labour market participation and social adulthood is far too narrow. The labour market is only one of many routes to social adulthood for young women and men, and this multiplicity of pathways needs to be reflected in both policy and public discourse.

As rural youth show great determination in their struggles to complete their education, they often combine schooling and paid work, and in many cases use the income from their work to cover the costs of their education. This avenue for educational progression is more open to young men who often have relatively more freedom to engage in paid work, whereas young women can be more confined in contexts of strongly patriarchal–gerontocratic institutions that lead them into early childbearing and marriage.

Finally, while a deficit model of African rural youth is at the core of dominant framings, there is also a counter tendency to reify the innovative and transformational capacity of young people. The former is as misplaced as the latter. Policy narratives that suggest young people are poised (or can be positioned) to transform agriculture and the food system more broadly, are unrealistic and counterproductive (Glover and Sumberg, 2020). While superficially these narratives might appear as a great vote of confidence in young people, they can also be seen as a discursive offloading of responsibility on to their young shoulders – to create their own jobs, to save the agricultural sector, to make the food system sustainable and to build the nation.

Implications for policy

The findings presented in this book generate a number of implications for the focus, content and targeting of policy. Here, we elaborate the key insights for education and some other key policy areas supporting the livelihoods of rural young people.

Clearly there is much room to improve education policy as it relates to rural SSA. For example, making primary and secondary education accessible to all remains an unfinished project, despite the great leap forward in primary enrolment resulting from concerted action to address the Millennium Development Goals (MDGs). The Sustainable Development Goals (SDGs) extend the ambition that education should be 'affordable' to all across the whole life course – including access to vocational and tertiary education. Gender equality is also a key element of these goals, and this is also far from being realized. The fact that inability to meet the costs of attending school continues to stop many rural young people from continuing their formal education is nothing less than a catastrophic policy failure. Indeed, unless it is addressed, the more recent focus on quality in education, and how it should be measured and improved, will do little for rural children or the next generations of young people. The findings also point to a demand for vocational and technical training. However, given the desire and need of many rural young people to combine further training with demands of ongoing economic and caring activities, this must be provided in flexible and/or part-time modes. More research may clarify the kinds of complementary interventions that would best enable rural young people to take advantage of existing or augmented educational resources.

The call for attention to quality education is fully justified and the broad aims of the SDGs are to be applauded. However, the metrics of quality proposed in the SDGs remain too narrowly focused on attainment. Student assessment data is notoriously problematic and importantly, such measures do not speak to the broader ambitions of the SDGs. These broader ambitions can only be addressed with a much clearer recognition of the historical, ideological and patriarchal underpinnings of education policy and practice, that result in the promotion of ideals of life, transition and work that are alien to most rural settings. The assumption of a linear trajectory from schooling into work, and the devaluing of unpaid and domestic work are prime examples. These underpinnings are also reflected in the discriminatory and disempowering dynamics that arise from the gendered landscape of education. Again, the focus on internationally comparable metrics of quality is misplaced as long as these underpinnings are left unexamined.

The research supports calls for a root-and-branch interrogation of school curricula with a focus on how they, and the whole schooling system, valorize (or denigrate) particular kinds of work and reproduce particular gender regimes. This interrogation should also address 'vocational' fields and locally relevant knowledge. Attention to differences in education systems that reflect different histories would also be valuable for understanding how the skills, vocational and employability agendas might be better integrated into mainstream schooling. Indeed, the content and skill specificity of formal education are already being questioned, with one example being the increasing investments in vocational skill training projects, outside mainstream schooling, aimed at young people (e.g. training of 'agripreneurs'). This is despite the fact that the evidence reviewed by Fox and Kaul (2018) 'casts serious doubt on the efficacy and value of training interventions to help youth enter formal wage employment'.

In relation to young people's work in agriculture, the findings assembled in this book offer solid support for the idea that in one way or another, large numbers of rural youth engage in crop and/or livestock production, and many combine this work with other economic activities. Further, both farming and livestock production have important places in the futures that many young people imagine for themselves – even in rural areas which may be seen as relatively undynamic. However, as noted above, in these futures these young people are not seeing or identifying themselves principally as farmers, or as having wholly agrarian livelihoods. Rather, theirs is often an arm's length, managerial or executive vision of engagement, with the manual work being done by hired labour, and farming being one of multiple economic activities. In effect, Africa's young people are developing their own unique take on the 'farming as a business' ideal that has been so heavily promoted over the last two decades. Specifically, they seem to be placing less emphasis on a model that assumes the key process underpinning any move to business-oriented farming will be the progressive specialization and professionalization of a hands-on 'farmer'. The young people's alternative vision, with their future selves as an (often) town-based manager at its centre, has more in common with the caricature of distant 'telephone farmers', directing farm operations through their digital devices. This vision poses important challenges in key areas of agricultural policy, including training and skills, employment, agricultural extension, and technology development and promotion. Here, young people may indeed be setting the agenda, and if so, it will be important that policy makers and programme designers remain flexible and in close touch with those on the forefront of change. This implies greater attention to who is able to access advisory services, both through existing systems and emerging digital variants. If the majority of young people engaged in farming are not household heads, then perhaps making agronomic extension more accessible to household dependents rather than exclusively to household heads would be advisable. This has implications for extension services that are tightly tied, for example, to cooperatives or other farmers' organizations, which may prioritize household heads or landlords because they are members.

The promotion of entrepreneurship is one of the preferred responses to the rural youth employment problem. However, across the agricultural commercialization hot and cold spots that were included in our study, there is relatively little evidence of young people engaging with value chains or 'farming as a business'. Nor is there much evidence that they are engaging in non-farm activities that require more than minimal levels of skill, investment or technology. This is not because they lack the capacity to do so, but because of the absence of rural consumers who need and can afford a more diverse and costly range of products and services. Further, the findings show that basic material factors such as the lack of roads and transport options hamper trade and mobility; whereas mobile money apps may go some way to facilitate financial transactions, they are certainly not available to or used by all. The critical assumption is that the structural conditions and opportunities are in place such that entrepreneurship – conceived as something more than simply low-level self-employment – makes sense. Yet it appears that this assumption cannot be sustained when it comes to the majority of rural areas. Entrepreneurship may be lucrative for a few young people in some places, but programmes that promote 'entrepreneurship for all' must now be designed with and accompanied by a large dose of honesty and realism.

More broadly, it is time to reflect critically on the increasingly common interventions that combine entrepreneurship training with 'financial inclusion'. This coupling is most often embedded in a three-part belief system – that agricultural value chains offer opportunity; that youth are innovative; and that digital technology is a rural game changer. Individually, these beliefs are either meaningless (e.g. youth are innovative) or they are appropriate only in some contexts and for some young people. As a general belief system or programming framework it has little value, yet as long as the focus is on small-scale, youth-targeted projects it will likely prove difficult to shift. In any case, more evidence on the impacts of tying training to financial interventions is needed (Fox and Kaul, 2018). The new framing discussed in the section above, with its focus on structural constraints to decent work, should help guide such research, and in so doing, open up space for a fundamental rethink of the training and skills agenda (Sumberg *et al.*, 2020).

A final policy area highlighted by the findings in this book is that of social protection. While rural residents generally are poorly covered by social protection programmes, young people may be particularly vulnerable. It is clear they face hazard as they go about building their livelihoods. These vulnerabilities are not associated with high-risk, high-return entrepreneurial endeavour, but rather everyday events like sickness (of themselves and family members), accidents, theft, business collapse because of customer non-payment, drought and so on. To recover from such events, individuals and young households may need to liquidate assets or use savings that cannot then be reinvested in farming or for starting and expanding new ventures. The findings show that the experience of hazard can also have major knock-on effects such as a child being withdrawn from school. There is an important opportunity to explore social protection interventions – beyond, for example, crop insurance – that could help protect young people, and all rural residents, against these downside risks. Such interventions could also, of course, have potentially strong synergistic effects with extension, education, training, and other investments targeting youth in the variable and risk-prone rural context that we find to be so pervasive.

Implications for research

Chapter 2 (this volume) and the research reported in the other chapters of *Youth and the Rural Economy in Africa: Hard Work and Hazard* amply illustrate the strengths and weaknesses of the frameworks, methods, data and analytical approaches that underpin current knowledge of young people's engagement with, and co-creation of, the rural economy. Here we highlight some specific steps with the potential to significantly increase the quality and relevance of the evidence base, and thus contribute positively to knowledge, policy and practice.

First and foremost there is a need, and an opportunity, to bring a broader set of perspectives to the discussion of rural youth livelihoods in SSA. For example, too much research, policy and public discourse, and youth-oriented development practice, does not draw on or engage with the large, diverse and challenging, yet highly relevant, literature from the field of youth studies. In his recent book *Agriculture and the Generation Problem*, Ben White (2020) demonstrates the benefits of integrating insights from youth studies.

While nationally representative household surveys provide valuable insights into the economic activities of rural young people, it is also recognized that these same household-based survey instruments may not fully capture youth activities. For example, do the farming activities of young people show up reliably in the plot rosters of the household? As discussed in Chapter 6 (this volume), the low level of plot management assigned to household dependents (other than spouse(s)) suggests that there may be systematic omissions. Two other questions deserve attention. First, how well are transitions (i.e. household formation, starting in farming, school-to-work, and migration) and path dependencies captured, and how might they be better interrogated in empirical work? Second, how can data be realistically collected on temporally and spatially variable livelihood engagement, including labour allocation to different activities, and income? This is not a new question, but despite having been raised over decades, it has still not been adequately addressed. A final concern (also raised in Chapter 6, this volume) is how well collective agency (to which youth contribute) is conceptualized and measured in these surveys.

A second, related area that deserves more attention is how the insights arising from qualitative research instruments can both be more creatively integrated with quantitative analyses, and more effectively inform policy. As highlighted in Chapter 2 (this volume) much of the qualitative research on youth in SSA relies on an extremely limited range of methods. Digging deeper into the qualitative toolbox, more methodological innovation, and new approaches to analysis and synthesis will likely be key to greater policy impact. For example, more creative mixing of quantitative and qualitative approaches may be used to better observe incipient change and innovation, and their social and technological dynamics. Along similar lines, new methodological approaches are needed to identify and sample sites of economic dynamism, i.e. areas undergoing rapid transformation, which may be atypical in a statistical sense, but can be very informative about how young people negotiate change.

Now we turn to three areas arising from the research reported in *Youth and the Rural Economy in Africa: Hard Work and Hazard* that deserve further attention:

- How do rural people – young and old, women and men – understand notions of work and decent work; and how do these understandings affect their engagement with school and training, present economic and domestic activities, and imagined futures? In particular, this question should explore the implications for young women of their significant burden of unpaid work, both domestic and reproductive, and the fact that, in many locations, they are expected to marry outside their natal homes.
- More research is needed around the nexus of home, school and work within rural SSA contexts. This should examine in more depth the processes of schooling, and pay attention to differences in curriculum and the gender- and class-based imaginaries of work that these reproduce. A related question is around the meaning of quality education to young women and men, and adults, and the gender dimensions of quality education.
- Third, while there are real differences between rural places, how do these differences matter for youth livelihoods and imagined futures? In fact, this was a central concern of our research, but our framework of hot spots and cold spots was apparently not strong enough to provide clear insights. The descriptive patterns of youth economic engagement and imagined futures were largely indistinguishable across hot versus cold areas. On the other hand, in some sites there were signs that social stratification was significant, and a stronger focus on how such stratification plays out in youth livelihood building would be valuable. Another tack is to think more carefully about how to define economic remoteness and dynamism. It is possible that the same criteria (e.g. population density, distance from markets) map on to very different realties in different countries, or that economic vibrancy is too complex to be captured by simple characterizations. More empirical work would help to clarify this.

Practice

In terms of development practice, *Youth and the Rural Economy in Africa: Hard Work and Hazard* points to two simple guidelines. First, as argued previously, practitioners need to be extremely cautious about youth-specific arguments and the youth-targeted interventions these are used to justify. While it is obviously true that 'youth are the future', *their* future is unlikely to improve through piecemeal interventions that support a small number of young people for a short period of time, without shifting opportunity structures. Focusing on opportunity structures and structural conditions requires programme continuity and coordinated, national and subnational approaches. Second, it is critical to work with, not against, the grain of family and social relations, as in most cases they allow young people to access key resources. This will also serve as a reminder that while interventions are often framed narrowly around economic activity and employment, young women and men build their livelihoods and move toward social adulthood through hard work on many fronts – including caring, relationships, education, children and civic action.

References

Abay, K., Asnake, W., Ayalew, H., Chamberlin, J. & Sumberg, J. 2020. Landscapes of opportunity: Patterns of young people's engagement with the rural economy in sub-Saharan Africa. *Journal of Development Studies (DOI: 10.1080/00220388.2020.1808195).*

Carreras, M., Sumberg, J. & Saha, A. 2020. Work and rural livelihoods: the micro dynamics of Africa's 'youth employment crisis'. *European Journal of Development Research* (doi.org/10.1057/s41287-020-00310-y).

Chamberlin, J. & Sumberg, J. 2020. Do youthful households farm differently? *Unpublished paper.*

Crossouard, B., Dunne, M., Szyp, C. & Flynn, J. 2020. Rural youth in Sub Saharan Africa: fractured lives and impossible futures. Unpublished paper.

Dejaeghere, J., Morris, E., & Bamattre, R. 2020. Moving beyond employment and earnings: reframing how youth livelihoods and wellbeing are evaluated in East Africa. *Journal of Youth Studies,* 23, 667–685.

Fox, L. & Kaul, U. 2018. *The Evidence Is In: How Should Youth Employment Programs in Low-Income Countries Be Designed? Policy Research Working Paper WPS 8500,* Washington, DC, The World Bank Group.

Glover, D. & Sumberg, J. 2020. Youth and Food Systems Transformation. *Frontiers in Sustainable Food Systems,* 4 (doi.org/10.3389/fsufs.2020.00101).

Oosterom, M., Flavia Namuggala, V. & Szyp, C. 2021. Family dynamics and implications for young people's 'struggles' in post-conflict northern Uganda. *Disasters* (in press).

Roberts, K. 1968. The entry into employment: an approach towards a general theory. *Sociological Review,* 16, 165–184.

Roberts, K. 1977. The social conditions, consequences and limitations of careers guidance. *British Journal of Guidance & Counselling,* 5, 1–9.

Roberts, K. 2009. Opportunity structures then and now. *Journal of Education and Work,* 22, 355–368.

Sumberg, J. & Hunt, S. 2019. Are African rural youth innovative? Claims, evidence and implications. *Journal of Rural Studies,* 69, 130–136.

Sumberg, J., Glover, D., Chamberlin, J., Flynn, J. & Johnson, V. 2019. *Landscapes of Rural Youth Opportunity. IFAD Research Series 47,* Rome, IFAD.

Sumberg, J., Flynn, J., Fox, L., Mader, P. & Oosterom, M. 2021. Africa's 'youth employment' crisis is actually a 'missing jobs' crisis. *Development Policy Review* (doi.org/10.1111/dpr.12528).

White, B. 2020. *Agriculture and the Generation Problem,* Black Point, Nova Scotia, Fernwood Publishing.

Yeboah, T., Chigumira, E., John, I., Anyidoho, N. A., Manyong, V., Flynn, J. & Sumberg, J. 2020. Hard work and hazard: Young people and agricultural commercialisation in Africa. *Journal of Rural Studies,* 76, 142–151.

Index

Note: The page numbers in bold and italics represents tables and figures.

CABI – who we are and what we do

This book is published by **CABI**, an international not-for-profit organisation that improves people's lives worldwide by providing information and applying scientific expertise to solve problems in agriculture and the environment.

CABI is also a global publisher producing key scientific publications, including world renowned databases, as well as compendia, books, ebooks and full text electronic resources. We publish content in a wide range of subject areas including: agriculture and crop science / animal and veterinary sciences / ecology and conservation / environmental science / horticulture and plant sciences / human health, food science and nutrition / international development / leisure and tourism.

The profits from CABI's publishing activities enable us to work with farming communities around the world, supporting them as they battle with poor soil, invasive species and pests and diseases, to improve their livelihoods and help provide food for an ever growing population.

CABI is an international intergovernmental organisation, and we gratefully acknowledge the core financial support from our member countries (and lead agencies) including:

UKaid from the British people

Ministry of Agriculture People's Republic of China

Australian Government Australian Centre for International Agricultural Research

Agriculture and Agri-Food Canada

Ministry of Foreign Affairs of the Netherlands

Schweizerische Eidgenossenschaft Confédération suisse Confederazione Svizzera Confederaziun svizra

Swiss Agency for Development and Cooperation SDC

Discover more

To read more about CABI's work, please visit: **www.cabi.org**

Browse our books at: **www.cabi.org/bookshop**,
or explore our online products at: **www.cabi.org/publishing-products**

Interested in writing for CABI? Find our author guidelines here:
www.cabi.org/publishing-products/information-for-authors/